D0854874

# MATHEMATICS:
## Art and Science

S. M. DOWDY

JOHN WILEY & SONS, INC.
New York London Sydney Toronto

Library of Congress Catalogue Card Number: 70-140516

ISBN 0-471-22020-5

Printed in the United States of America

10 9 8 7 6 5 4 3 2 1

*For Tom and Bob*

# Preface

**A** world-famous mathematician once gave a lecture entitled "What does it mean to be an algebraist?" The talk was about some recent developments he had made in algebra. The outstanding thing about the speaker was his way of communicating his enthusiasm and appreciation for his subject to his audience. He concluded his lecture by saying, "I hope you understand now that to an algebraist, algebra is a way of life." I am sure that most mathematicians would agree to a similar statement: "To a mathematician, mathematics is a way of life."

I find it very difficult to be quiet about what I consider to be my "way of life." I wish that others could enjoy the thrill of mathematical discovery, see the beauty in a proof, and understand the role mathematics has played in our culture. I believe that it is possible for most people to have this appreciation, even if their interests and abilities are not particularly mathematical. In the light of my views, it is not surprising that, along with classes for mathematics majors at Oklahoma State University, I began to teach a course in mathematics for liberal arts students. This book is the outgrowth of these classes.

I have written the book as a college text for nonmathematicians and nonusers of mathematics who have had at least one year of high school algebra. A year of high school geometry would also help readers, but it is not essential. My purpose is to develop an appreciation of mathematics and an awareness of its contribution to our culture. I have *not* attempted to present the specialized material needed by elementary mathematics teachers, beginning mathematics majors, or students of any of the natural and physical sciences. (My experience has been, however, that this second group finds the material, taken as an elective course, both profitable and enjoyable. The material brings a unity and insight to mathematics that they usually do not see in their required courses.)

I believe that the best way to show nonmathematicians the beauty and elegance of mathematics is to guide them in a few in-depth experiences in mathematics. It is not important to show them all of the basic concepts of some area, nor always to use standard pedagogical approaches, notation, or definitions. What is important is that the nonmathematicians follow the thought pattern of the working mathematician: experimentation, intuition, conjecture and, finally, proof or disproof. Once a topic is completed, I think it is important to step back and discuss what that topic illustrates about the nature of mathematics, what branch of mathematics the topic belongs to, and to talk a little about the history and applications of that branch.

This book is my way of carrying out this program. Hopefully I will be forgiven for giving only brief glimpses at the history and applications of mathematics. Since these unavoidable deficiencies can and should be offset by outside reading, I have included references to many of the excellent related expository articles and books. I have tried to avoid references to textbooks and technical material. Most of the readings are specifically written for nonmathematicians. I have included under the heading "Suggested Readings" four collections that contain many of the specific references given throughout the book.

S. M. Dowdy

Morgantown, West Virginia
1970

# Note to the Instructor

The text begins with a short introduction that poses the question: What is mathematics? Nine chapters follow, each one developing in depth one significant topic from a branch of mathematics. The nine chapters include more material than is needed in a three-hour one-semester course (in such a course I can cover only four or five chapters).

The following summary of the content, background needed, and purpose of each chapter will help the instructor to decide which of the chapters to study.

Chapter 1 Number Theory

*Covers:* $d(N)$, the number of divisors of a number; its properties.
*Significant problem:* the proof of the formula for $d(N)$.
*Background needed:* multiplication and division of natural numbers.
*Teaches:* the role of intuition in mathematics; the meaning of proof and counterexample; proof by contradiction. Discusses the specialized symbolic language of mathematics and some open questions in number theory.

Chapter 2 Modern Algebra

*Covers:* groups; subgroups; normal subgroups (optional).
*Significant problem:* the Lagrange Theorem.
*Background needed:* notion of set and subset; definition of divisor.
*Teaches:* how a mathematician introduces a new tool to solve a problem; how mathematics moves from the concrete to the abstract and then sometimes back to the concrete. Discusses mathematics as an art; briefly surveys modern algebra.

Chapter 3 Geometry

*Covers:* projective geometry.
*Significant problem:* a proof that Pappus' Theorem implies Desargues' Theorem.

*Background needed:* knowledge of basic geometrical terms, such as point, line, and triangle.
*Teaches:* that one part of mathematics can be approached by different methods. Short discussion of axiom systems, non-Euclidean geometries, and applications of non-Euclidean geometries.

Chapter 4 Foundations

*Covers:* lattices.
*Significant problem:* a proof of the associativity of *cup* in a lattice.
*Background needed:* at least two of Chapters 1, 2, and 3, especially Section 6 of each.
*Teaches:* the role of axiom systems in mathematics; what they are, their economy, what they cannot do (Gödel's Incompleteness Theorem).

Chapter 5 Mathematical Logic

*Covers:* two- and three-valued logic of sentences.
*Significant problem:* axioms for logic; truth tables.
*Background needed:* some previous exposure to mathematical thinking (this chapter should not be studied first).
*Teaches:* that the subject matter of mathematics can be things other than numbers and geometrical figures; an illustration of generalization. Surveys some new areas of mathematics: game theory, automata, communications theory, cybernetics, and mathematical linguistics.

Chapter 6 Set Theory

*Covers:* infinite sets; cardinality; cardinal numbers.
*Significant problem:* Cantor's Theorem about the power set of a set.
*Background needed:* notion of set, element of a set, and subset. (Some number systems and binary notation appear in this chapter, but with enough explanation that a strong background is not presumed.)
*Teaches:* the unintuitive is often true in mathematics; seemingly small assumptions often lead to powerful theorems. Discusses Russell's Paradox.

Chapter 7 Analysis

*Covers:* infinite series; notion of limit.
*Significant problem:* convergence of series according to the first arithmetic mean as a generalization of ordinary convergence.

*Background needed:* strong skill with fractions and basic algebraic manipulations (a class without this background will find this chapter difficult).
*Teaches:* how a mathematician generalizes; some criteria for acceptable definitions; how mathematics can represent physical change.

Chapter 8 Probability

*Covers:* basic properties of probability.
*Significant problem:* Hardy's Law for population genetics.
*Background needed:* notion of set; vague acquaintance with Mendel's First Law of Genetics.
*Teaches:* the method of applied mathematics. Discusses pure versus applied mathematics.

Chapter 9 Computer Science

*Covers:* what a computer is.
*Significant problem:* flow diagrams; repetitive processes.
*Background needed:* nothing specific.
*Teaches:* what a computer can and cannot do; the relation of a computer to mathematics; what a computer may be able to do in the future.

If Chapter 4 is not covered, Section 6 of Chapters 1, 2, and 3 can be (and probably should be) omitted since their only purpose is to provide background for the discussion of axiom systems in Chapter 4. With the exception of Chapter 4, which should be preceded by at least two of Chapters 1, 2, and 3, any order can be followed. Most classes will find Chapters 5 and 7 more difficult than the rest.

S.M.D.

# Acknowledgments

I thank all the people who contributed to this book in one way or another. I am particularly grateful to my students at Oklahoma State University for their criticisms and reactions to the preliminary form of the text; to Leroy Folks of Oklahoma State University for his interest and remarks, which led to basing Chapter 8 on Hardy's Law; to Al Johnson of West Virginia University for class testing several chapters and for his helpful suggestions; and to W. Newman Bradshaw of West Virginia University for checking the biological facts in Chapter 8.

S.M.D

# Suggested Readings

Specific suggestions for readings are given throughout this book. The majority of the readings can be found in the following four collections:

*Mathematics in the Modern World,* Readings from *Scientific American* with Introduction by Morris Kline, W. H. Freeman, San Francisco, 1968 (available in paperback). In the references any *Scientific American* article preceded by an asterisk is also in *Mathematics in the Modern World.*

*The Mathematical Sciences,* A Collection of Essays, Edited by the Committee on Support of Research in the Mathematical Sciences with collaboration of George A. W. Boehm, The M.I.T. Press, Cambridge, Mass., 1969 (available in paperback).

*The World of Mathematics,* four volumes, Edited by James R. Newman, Simon and Schuster, New York, 1956 (available in paperback).

The current edition of *Encyclopaedia Britannica.* (A complete revision was done in 1966; articles in editions from then on are up to date. These articles are usually more technical than those in the preceding three references.)

# Contents

# MATHEMATICS

# INTRODUCTION

# INTRODUCTION

## What Is Mathematics?

If asked, "What is mathematics?" most of us would be caught off guard. We might make one of the following comments:

"It's abstract."
"Works with numbers."
"Solves problems."
"Proves theorems about figures."
"Has something to do with sets."
"Has rules for getting answers."
"Uses equations."
"Adds and subtracts."
"Studies known facts, never finds anything new."

All of these remarks, except the last, would be correct, but none of them would define mathematics.

Our inability to define mathematics should not embarrass us, nor should it increase our fear that mathematics is some erudite subject which we will never be able to understand. Actually, no one, not even the greatest mathematician, has given a completely satisfactory definition of mathematics. The attempts are usually somewhat philosophical and not very helpful. The late Bertrand Russell said,

> ... mathematics may be defined as the subject in which we never know what we are talking about, nor whether what we are saying is true.

Charles Steinmetz explains this another way,

> Mathematics is the most exact science, and its conclusions are capable of absolute proof. But this is so only because mathematics does not *attempt* to draw absolute conclusions. All mathematical truths are relative, conditional.

Benjamin Peirce was a little more direct,

> Mathematics is the science which draws necessary conclusions.

His son, Charles Peirce, elaborated on this,

> Mathematics is the study of what is true of hypothetical states of things. That is its essence and definition.

Why is it so difficult to define mathematics? If anything is definite, we feel that mathematics is. It should have a definition. Maybe the trouble is not with mathematics, but with ourselves for expecting a simple answer to this question. Have we ever seen a simple definition of music, poetry, history—or for that matter, of friendship, hate, love? Many of the most familiar things in our lives defy definition. The area is too rich or too complex to be captured in twenty-five words or less. We do not define music. We listen to it, learn how it is composed, play it, perhaps compose some ourselves. We have no doubt about what music is, but we can't define it. So with mathematics. If we are going to understand what it is, we must see some of it, learn how it is done, discover some of it ourselves.

We can learn to appreciate music just by listening. The person who plays an instrument, however, appreciates it more. The one who plays with a group understands it best. In mathematics the same thing is true. We can learn a little about the nature of mathematics by reading about it or listening to someone lecture on it. A real understanding, however, is gained only if we do some of it ourselves. We would understand it best if we could talk with others about mathematical ideas.

Since this is the case, our plan is clear. To answer the question, "What is mathematics?" we must *do some* and learn how to talk about it. But what kind of mathematics? There are many branches of mathematics, and besides most problems would require a great deal of mathematical background. The number of branches is not a problem, in fact it will lend variety to our study. We'll try to do something from most of the major branches. The problem of background is a little more difficult. Although many problems require years of technical preparation, there are problems in every branch that most people can understand with a little bit of explanation. We'll do some of these. After we have done some number theory, modern algebra, modern geometry, foundations, logic, set theory, analysis, probability, and computer science, we'll look once again at the question "What is mathematics?"

## Exercises

1. In order to see how mathematicians communicate with each other, go to the library and browse for half an hour among the mathematical periodicals and the mathematics books. How many journals are devoted exclusively to mathematics? What are some of the fields covered by the books and journals? Has this trip to the library changed your impression of mathematics in any way? If so, how?
2. Read from one of the following and describe any new ideas it gives you about the nature of mathematics. (You may find the material a little difficult. Do not

let this discourage you. If it is difficult now, try to read it again after you have read this book.)

Bell, E. T., *Men of Mathematics,* Simon and Schuster, New York, 1937.

Bergamini, David, and the Editors of Life, *Mathematics,* Life Science Library, Time Incorporated, New York, 1963.

Bourbaki, N., "The Architecture of Mathematics," *American Mathematical Monthly,* Vol. 57, 1950, pp. 221–232.

*Courant, R., "Mathematics in the Modern World," *Scientific American,* Vol. 211, No. 3, September, 1964, pp. 41–49.

Halmos, P. R., "Mathematics as a Creative Art," *American Scientist,* Vol. 56, No. 4, 1968, pp. 375–389.

*Halmos, P. R., "Nicolas Bourbaki," *Scientific American,* Vol. 196, No. 5, May, 1957, pp. 88–99.

Russell, Bertrand, "My Mental Development," in *The World of Mathematics,* Vol. 1, J. R. Newman, Simon and Schuster, New York, 1956, pp. 381–394.

Turnbull, Herbert W., "The Great Mathematicians," in *The World of Mathematics,* Vol. 1, J. R. Newman, Simon and Schuster, New York, 1956, pp. 75–168.

* *Scientific American* articles preceded by an asterisk are also in *Mathematics in the Modern World,* Readings from *Scientific American* with Introduction by Morris Kline, W. H. Freeman, San Francisco, 1968.

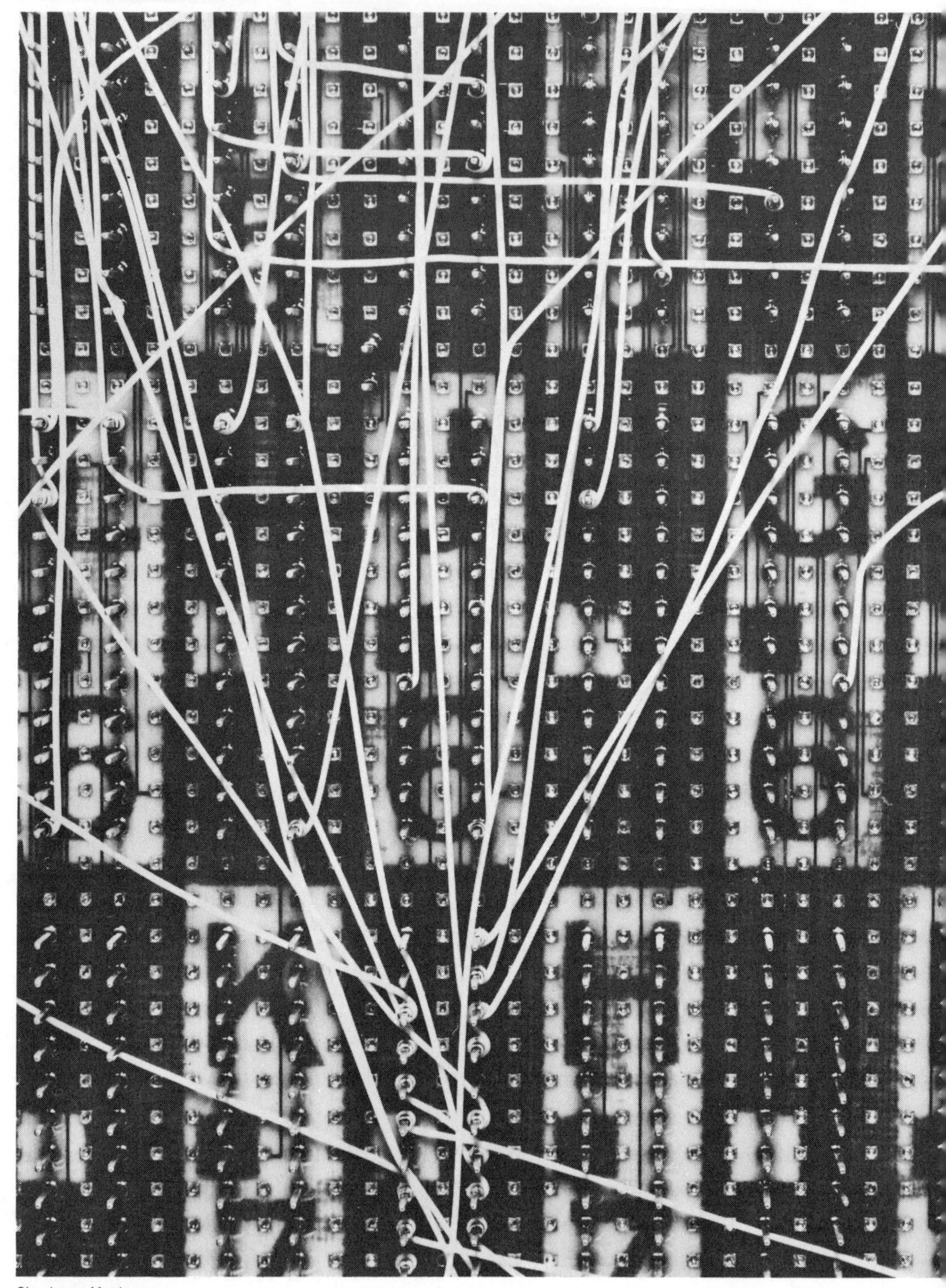
Shackman-Monkmeyer

1

# NUMBER THEORY

## Discovering New Properties Among Old Numbers

**Binary arithmetic, studied long ago in number theory, is now used in computers.**

**S**ince most of us think that mathematics is about numbers and calculating, let's begin with some number theory. We will see that, although it is true that mathematicians sometimes study numbers, they are rarely concerned with calculating.

### 1. THE DIVISORS OF A NUMBER

Let's imagine that we are planning to tile a floor which measures 12 by 15 feet. We want to use only whole tiles and we want all of them to be the same size. If tiles are available in 4-, 5-, 6-, 8-, 9-, and 12-inch squares, which sizes of tiles could we use?

***Stop and figure this out yourself.***

After a little bit of calculating we should reply that we could pick from the 4-, 6-, 9-, or 12-inch tiles, but that neither the 5- nor 8-inch tiles would work. This is because a 12 x 15 foot floor is 144 by 180 inches, and 4, 6, 9, and 12 all divide into 144 and 180 leaving no remainder. There is a remainder if we divide 144 by 5 and if we divide 180 by 8. Let's formalize these notions.

Mathematicians say that the natural number $K$ *divides* the natural number $N$ if there is no remainder when $N$ is divided by $K$. (The *natural numbers* are the ones we count with, 1, 2, 3, . . . .) It is also possible to define the notion "$K$ divides $N$" without using division: A natural number $K$ divides the natural number $N$ if there is a natural number $T$ such that $K \cdot T = N$. For example, 4 divides 144 since $4 \cdot 36 = 144$. These two definitions are the same, of course, since $N$ divided by $K$ is $T$ if, and only if, $K \cdot T = N$.

We were fortunate that our floor measured 144 by 180 inches. If the dimensions had been 137 by 173 inches, we could not tile it as we wanted because none of the numbers, 4, 5, 6, 8, 9, or 12 divides either 137 or 173. The only natural numbers that divide 137 are 1 and 137. Only 1 and 173 divide 173.

It seems that some natural numbers have several natural numbers which are divisors and others have only a few. Let's find the number of divisors of some natural numbers. Let $N$ be any natural number and let's write

$d(N)$ for the *number of natural numbers which are divisors* of $N$ (we read this "*d* of *N*")*.*

$$\text{If } N = 1,\ d(N) = 1;$$
$$\text{if } N = 2,\ d(N) = 2;$$
$$\text{if } N = 3,\ d(N) = 2.$$

***Complete the following table for* N *from 1 to 25. (To make sure your completed table is correct, compare it with the table of one of your friends.)***

| $N$ | Divisors of $N$ | $d(N)$ |
|---|---|---|
| 1 | 1 | 1 |
| 2 | 1, 2 | 2 |
| 3 | 1, 3 | 2 |
| 4 | 1, 2, 4 | 3 |
| . | . . . | . |
| . | . . . | . |
| . | . . . | . |
| 25 | 1, 5, 25 | 3 |

If we had wanted a table from 1 to 1000 we would soon get tired of all the calculations and begin to wonder if there is a shorter way to find $d(N)$. Maybe we can find a clue to this short method if we look at the table we have just completed.

***Using your completed table see if you can find any relationship between each* N *and its corresponding* d(N).**

We might notice that there are several 2s in the $d(N)$ column: $d(3) = 2$, $d(5) = 2$, $d(11) = 2$, $d(23) = 2$. In each case, $N$ is an odd number. Maybe $d(N)$ is 2 if $N$ is odd. Let's call this guess a *conjecture.*

CONJECTURE: If $N$ is odd, then $d(N) = 2$.

From our table we have some evidence that this conjecture might be true, then again maybe we have jumped to this conclusion too quickly.

* The usual notation for $d(N)$ is $\tau(N)$, read "tau of $N$." Since Greek letters sometimes cause a psychological block to learning for those not familiar with the Greek alphabet, we will not use them.

What we must do now is either prove that this conjecture is true, or prove that it is false. To show that it is true we would have to reason from some basic facts of arithmetic; to show that it is false all we have to do is to find a single example for which it is not true. Such an example is called a *counterexample.*

---

***Can you find a counterexample to the conjecture?***

---

Luckily, we don't have to look far, $d(9) = 3$ and 9 is odd, so if $N$ is odd, $d(N)$ is not necessarily 2. We would also be correct if we had given as a counterexample $d(1) = 1$, $d(15) = 4$, $d(21) = 4$, or $d(25) = 3$.

Since we have found a counterexample to our conjecture, our conjecture is false. We have not yet found a relationship between each $N$ and the corresponding $d(N)$. We will take a harder look at our table in the next section.

## Exercises

1. List all of the divisors of $N$ for $26 \leq N \leq 50$ (for all $N$ from 26 to 50 including 26 and 50).
2. What is $d(N)$ for $26 \leq N \leq 50$?
3. Make a graph plotting $d(N)$ against $N$ for $0 \leq N \leq 50$.
4. Why is a single counterexample sufficient to show that a conjecture is false?
5. Is $d(2) = 2$ a counterexample to our conjecture? Why, or why not?
6. Find a counterexample for the following statements:
   (a) The tallest man in the world is seven feet tall.
   (b) The sum of two odd numbers is an odd number.
   (c) Fractions can be added by adding the numerators and the denominators.
   (d) All right triangles are similar.
   (e) For all integers $X$, $X^2 > 0$. (The integers are 1, 2, 3, . . . , −1, −2, −3, . . . and 0. The notation $X^2 > 0$ means "$X^2$ is greater than 0.")

### 2. THE PRIME FACTORIZATION OF A NUMBER

Let's look at our table again to see if we can find a quick way to get $d(N)$ for any natural number $N$ if $N$ is greater than 1. We found that $d(N) = 2$

does not mean that $N$ is odd. For what values of $N$ is $d(N) = 2$? If $N$ is 2, 3, 5, 7, 11, 13, 17, 19, or 23, we notice that $d(N) = 2$.

---

***What kind of numbers are 2, 3, 5, 7, 11, 13, 17, 19, and 23?***

---

The numbers 2, 3, 5, 7, 11, 13, 17, 19, and 23 are all primes. A *prime number* is a natural number greater than 1 which has only 1 and itself as divisors. This is a clue. Maybe $d(N)$ has something to do with primes.

If a number greater than 1 is not prime, then we say it is *composite.* For example, 6 is not prime because it has 2 as a divisor; 6 is composite. Other examples of composite numbers are 12, 27, 44, and 100. (The number 1 is neither prime nor composite.) Since we find $d(N)$ for all the natural numbers including the composite numbers and $d(N)$ is somehow related to primes, we'll have to find out how the composite numbers are related to the primes.

The characteristic thing about composite numbers is that they have divisors other than themselves and 1. The composite 24, for example, has 1, 2, 3, 4, 6, 8, 12, and 24 as divisors. We could represent 24 several different ways by using these divisors as factors in a product which equals 24. For example:

$$24 = 2 \cdot 12$$
$$24 = 3 \cdot 8$$
$$24 = 1 \cdot 24$$
$$24 = 6 \cdot 4$$
$$24 = 2 \cdot 6 \cdot 2$$

are some of the ways we could represent 24.

---

***Can 24 be represented as a product in which all of the factors are primes?***

---

The answer is yes, and one systematic way to find the factorization is to begin with the smallest prime divisor of 24 and use it as a factor. The smallest prime divisor is 2, so we can write

$$24 = 2 \cdot 12.$$

Now we look at 12 and find its smallest prime divisor; again this is 2, so we write

$$24 = 2 \cdot 2 \cdot 6.$$

Looking at 6, it also has a prime divisor 2, and we can write

$$24 = 2 \cdot 2 \cdot 2 \cdot 3.$$

The number 3 is prime so we cannot carry the process any further.

---

***Represent each of the following numbers as a product of prime factors: 15, 36, 64, and 126.***

---

We should answer that $15 = 3 \cdot 5$, $36 = 2 \cdot 2 \cdot 3 \cdot 3$, $64 = 2 \cdot 2 \cdot 2 \cdot 2 \cdot 2 \cdot 2$, and $126 = 2 \cdot 3 \cdot 3 \cdot 7$. Sometimes a prime appears as a factor only once, sometimes several times. For some composites the factors are different, for others they are all the same.

The method described above for finding a prime factorization is one way to approach the problem. There are other approaches and we should be wondering at this point if they might lead to different prime factorizations.

---

***Use a different method to find the prime factorization of 24. Do you get a factorization different than $24 = 2 \cdot 2 \cdot 2 \cdot 3$?***

---

Another method would be to begin with any factorization (not necessarily prime) of 24, as $24 = 6 \cdot 4$. We can then replace each of these factors by some product which they equal. For example, $6 = 2 \cdot 3$ and $4 = 2 \cdot 2$, so we write $24 = 2 \cdot 3 \cdot 2 \cdot 2$. Sometimes it will take several repetitions of this process of replacing a factor by a product before we reach primes exclusively. It is not a particularly systematic way to approach the problem, but it is an approach which works. Notice that this second approach gave us $24 = 2 \cdot 3 \cdot 2 \cdot 2$. Although the factors are arranged in a different order, this is still essentially the same factorization as $24 = 2 \cdot 2 \cdot 2 \cdot 3$, because the same primes appear the same number of times.

It seems that 24 can be written as a product of primes in only one way (disregarding the arrangement of the factors). Not only 24, but every composite number seems to have the property that it can be written as a product of primes in only one way. If we can do something in only one way, we call it a *unique* way.

The observation which we have just made is in fact true for all composite numbers and is called the *Fundamental Theorem of Arithmetic.*

THE FUNDAMENTAL THEOREM OF ARITHMETIC: Every composite number can be written as a product of primes in a unique way.

If we were studying number theory thoroughly we would stop and prove this theorem. At first it looks obviously true, but it is possible to invent a number system in which it is not true.* Since we are not attempting to learn all of number theory, but are only studying divisors in order to learn something about the nature of mathematics, we will assume that the Fundamental Theorem of Arithmetic is true.

We still have not found a quick way to find $d(N)$. We are, however, a little closer. Finding $d(N)$ quickly has something to do with primes, and every natural number greater than 1 is either a prime or can be written in a unique way as a product of primes. In Section 3 we will carry our search further.

## Exercises

**1.** List all of the primes between 1 and 50.

**2.** Which of the following are composite numbers: 1, 15, 79, 92, 93, 101, 121, 135, 137?

**3.** What is the prime factorization of 30, 38, 46, 50, 80, 96, 144?

**4.** Represent 10 as the sum of two primes. Is this representation unique, that is, can 10 be represented as a sum of two primes in a different way? Represent 18 as the sum of two primes. Is this representation unique?

**5.** Look up S. K. Stein's description of a number system in which the Fundamental Theorem of Arithmetic is not true. Explain it to someone else.

### 3. A FORMULA FOR *d(N)*, THE NUMBER OF DIVISORS OF *N*

Let's recall where we are in our attempt to find an easy way to calculate $d(N)$. We could say now that *every* natural number greater than 1 is related to a prime or some primes in a unique way. If the natural number is a prime, it is related uniquely to itself. If the number is composite it is related uniquely to the primes in its prime factorization.

* See Sherman K. Stein, *Mathematics, The Man-Made Universe,* Second Edition, W. H. Freeman, San Francisco, 1969, pp. 49–51.

When writing out the prime factorization of a number we can use exponents to shorten the list of factors. Instead of writing $24 = 2 \cdot 2 \cdot 2 \cdot 3$ we can write $24 = 2^3 \cdot 3$.

---

***Complete the following list which indicates the prime factorization of each N from 2 to 25, inclusive.***

---

$$
\begin{array}{l}
2 = 2^1 \\
3 = 3^1 \\
4 = 2^2 \\
\ldots \\
\ldots \\
\ldots \\
12 = 2^2 \cdot 3^1 \\
\ldots \\
\ldots \\
\ldots \\
25 = 5^2
\end{array}
$$

---

***Using your table for* d(N) *from Section 1 and this last list, can you find any relationship between* d(N) *and the primes related to* N?**

---

We might notice that

$$
\begin{array}{rcl}
4 = 2^2 & \text{and} & d(4) = 3, \\
9 = 3^2 & \text{and} & d(9) = 3, \\
25 = 5^2 & \text{and} & d(25) = 3.
\end{array}
$$

In each case $d(N)$ is 3 and $3 = 2 + 1$, one greater than 2, the exponent of the primes. A similar thing happens if $N$ is a prime; for example, $d(5) = 2$ and $2 = 1 + 1$, one greater than the exponent in $5^1$.

---

***Can you find a similar relationship between $6 = 2^1 \cdot 3^1$ and d(6) = 4?***

---

We notice that $4 = 2 \cdot 2 = (1 + 1)(1 + 1)$, so that there is a factorization of $d(6)$ in which each factor is one greater than the exponent in the prime factorization of 6. Similarly, $12 = 2^2 \cdot 3^1$ and $d(12) = 6 = 3 \cdot 2 = (2 + 1)(1 + 1)$.

***Use the above method to find the values of* d(33), d(36), d(37), d(42), *and* d(48).**

We should answer that

$33 = 3^1 \cdot 11^1$, so it seems that $d(33) = (1+1)(1+1) = 4$,
$36 = 2^2 \cdot 3^2$, so it seems that $d(36) = (2+1)(2+1) = 9$,
$37 = 37^1$, so it seems that $d(37) = (1+1) = 2$,
$42 = 2^1 \cdot 3^1 \cdot 7^1$, so it seems that $d(42) = (1+1)(1+1)(1+1) = 8$,
and $48 = 2^4 \cdot 3^1$, so it seems that $d(48) = (4+1)(1+1) = 10$.

***Check that these answers are those you would have obtained had you found all the divisors and counted them. (Use your results from Exercise 2, Section 1.)***

Now we are almost ready to state a conjecture about how to calculate $d(N)$ quickly. There is only one difficulty. Even though we have an idea about what the relationship is between $N$ and $d(N)$, how are we going to put it into words? We will have to talk about the prime factorization of *any* number $N$ greater than 1. If we are not speaking of a specific $N$, how can we write out its prime factorization? Mathematicians have developed a specialized symbolic language to help them be precise even when speaking in general terms. We will use this language.

If $N$ is any number we can represent its prime factorization the following way: $N = p_1^{a_1} \cdot p_2^{a_2} \cdot \ldots \cdot p_n^{a_n}$ where each $p$ is a different prime and each $a$ is greater than 0. The three dots represent any primes we have not written in this representation of the factorization. There may, of course, be less than three primes in the factorization. This notation is not very difficult to understand. If $N = 60$, then $60 = 2^2 \cdot 3^1 \cdot 5^1$ and $p_1 = 2$, $p_2 = 3$, $p_3 = p_n = 5$, $a_1 = 2$, $a_2 = 1$, $a_3 = a_n = 1$. Although it is not essential to our reasoning, we will always write the primes in the prime factorization according to size beginning with the smallest.

***Write out the prime factorization of* N = 1176 *and* N = 420*. What are* n, $p_1$, $p_2$, *. . . ,* $p_n$, $a_1$, $a_2$, *. . . ,* $a_n$*?***

If $N = 1176 = 2^3 \cdot 3^1 \cdot 7^2$, then $n = 3$, $p_1 = 2$, $p_2 = 3$, $p_3 = p_n = 7$, $a_1 = 3$, $a_2 = 1$, and $a_3 = a_n = 2$. If $N = 420 = 2^2 \cdot 3^1 \cdot 5^1 \cdot 7^1$, then $n = 4$, $p_1 = 2$, $p_2 = 3$, $p_3 = 5$, $p_4 = p_n = 7$, $a_1 = 2$, $a_2 = 1$, $a_3 = 1$, and $a_4 = a_n = 1$.

With these symbols we are ready to state the general conjecture about how to calculate $d(N)$. This time the conjecture is true and since we can prove it, we'll call it a *theorem.*

> THEOREM: If the prime factorization of $N$ is $p_1^{a_1} \cdot p_2^{a_2} \cdot \ldots \cdot p_n^{a_n}$, then $d(N) = (a_1 + 1)(a_2 + 1) \ldots (a_n + 1)$.

---

***Calculate d(29), d(34), d(45), and d(49) using the theorem, and then check to see if these are the correct answers.***

---

The formula in the theorem seems to work for every number we try. However, this does not mean that the formula is correct. We may not have tried enough examples to uncover a case for which it fails to work. No matter how many specific examples we do, we have not proved the theorem. To make sure the theorem is true, we must reason from some basic facts of arithmetic and show that the formula works for *every* natural number greater than 1. We will do this in the next section.

## Exercises

1. Write out the prime factorization of $N$ for $26 \leq N \leq 50$.
2. Use the theorem to calculate $d(N)$ for 16, 21, 24, 26, 30, 32, 40, 44, and 50. Compare your answers with the values of $d(N)$ you found in the first section.
3. The prime factorization of any number $N$ can be represented by $N = p_1^{a_1} \cdot p_2^{a_2} \cdot \ldots \cdot p_n^{a_n}$. If $N$ is 30, what are $n, p_1, p_2, \ldots, p_n, a_1, a_2, \ldots$, and $a_n$? Do the same thing for 4, 15, and 180.
4. *Conjecture:* $2^p - 1$ is a prime if $p$ is a prime. Show that this conjecture is true for $1 < p < 11$. Is this a proof that the conjecture is true? Why, or why not? Prove the conjecture is true, or show that it is false by a counterexample, whichever case is correct.

## 4. A PROOF OF THE FORMULA FOR *d(N)*

As a preliminary to the proof of the formula for $d(N)$ we'll have to look at the prime factorization of each of the divisors of $N$ which is greater than 1. It seems that the prime factorization of a divisor of $N$ should be related to the prime factorization of $N$ itself.

***Write out the prime factorization of each divisor of 12 which is greater than 1 and compare it with the prime factorization of 12 itself.***

We should write

$$\begin{aligned} 2 &= 2^1 \\ 3 &= 3^1 \\ 4 &= 2^2 \\ 6 &= 2^1 \cdot 3^1 \\ 12 &= 2^2 \cdot 3^1. \end{aligned}$$

The relationship will become clear if we agree to let $K^0 = 1$ for any natural number $K$: then $2^0 = 1$ and $3^0 = 1$. We can then put the divisor 1 in the list and rewrite the list as follows:

$$\begin{aligned} 1 &= 2^0 \cdot 3^0 \\ 2 &= 2^1 \cdot 3^0 \\ 3 &= 2^0 \cdot 3^1 \\ 4 &= 2^2 \cdot 3^0 \\ 6 &= 2^1 \cdot 3^1 \\ 12 &= 2^2 \cdot 3^1 \end{aligned}$$

Now we can see that there is a factorization of each divisor of 12 which contains the same primes as 12 and in which the exponent of each prime is some integer from 0 up to the corresponding exponent in 12 inclusive.

***Write out the prime factorization of each divisor of 90 which is greater than 1 and see if a similar use of primes occurs.***

It is true that all the divisors of 90 have factorizations which contain the same primes as 90 and with each exponent some integer from 0 up to the corresponding exponent in the prime factorization of 90. This use of primes also occurs for *any* number $N$ and its divisors. We will prove this next. Since this fact will help us prove the theorem, we will call it a *lemma.* A lemma is a theorem which is needed to prove another theorem.

In the lemma we will look at the *prime factorization* of any divisor of $N$, that is, before we have added any primes with zero exponents. In the prime factorization each exponent will be some integer between 1 and the corresponding $a$.

LEMMA: If $N$ has prime factorization $p_1^{a_1} \cdot p_2^{a_2} \cdot \ldots \cdot p_n^{a_n}$ and $D$ not equal to 1 is a divisor of $N$ with prime factorization $q_1^{b_1} \cdot q_2^{b_2} \cdot \ldots \cdot q_m^{b_m}$, then every $q$ is a $p$ and its exponent $b$ is some integer from 1 up through the corresponding exponent $a$ of $p$.

*Proof of Lemma:* To prove that every $q$ is a $p$ we will show that if some $q$ isn't a $p$, then we can reason to something false. (This type of argument is called an indirect proof, or a *proof by contradiction.)*

Let $D$ be any divisor of $N$ and assume $D$ has a prime factor $q$ which is not a $p$. We can write $D = q \cdot R$ where $R$ is the product of all of the other prime factors of $D$. (For example, if $D = 60 = 2^2 \cdot 3 \cdot 5$ and $q = 2$, then $D = 2 \cdot 30$.) Since $D$ divides $N$, there is a natural number $T$ such that $N = D \cdot T$. Substituting $q \cdot R$ in place of $D$, we get $N = q \cdot R \cdot T$. Now we can factor $R \cdot T$ into its prime factors and we have a prime factorization of $N$ containing $q$. This must be false because we assumed that $q$ was not in the prime factorization of $N$ (it was not a $p$), and according to the Fundamental Theorem of Arithmetic there is only one prime factorization of $N$, $N = p_1^{a_1} \cdot p_2^{a_2} \cdot \ldots \cdot p_n^{a_n}$.

It remains now to prove that the exponent of each $q$ is some integer between 1 and the corresponding exponent $a$ of $p$ inclusive. This can be done indirectly in a way similar to the argument we just used and will be left as an exercise. ■

(We'll use a dark square ■ to indicate we have finished proving something.)

With this lemma we are now ready for a proof of the theorem about the formula for $d(N)$. Let's state it again so we know what we are trying to prove.

THEOREM: If the prime factorization of $N$ is $p_1^{a_1} \cdot p_2^{a_2} \cdot \ldots \cdot p_n^{a_n}$, then $d(N) = (a_1 + 1)(a_2 + 1) \ldots (a_n + 1)$.

Before we do the formal proof let's look at a way to get all of the divisors of a number $N$. Let $N = 60 = 2^2 \cdot 3^1 \cdot 5^1$. To form divisors we could use $2^0$, $2^1$, or $2^2$, $3^0$ or $3^1$, and $5^0$ or $5^1$ as factors. If we write out all combinations in a systematic way we get:

$$\begin{array}{lll} 2^0 \cdot 3^0 \cdot 5^0 & 2^1 \cdot 3^0 \cdot 5^0 & 2^2 \cdot 3^0 \cdot 5^0 \\ 2^0 \cdot 3^0 \cdot 5^1 & 2^1 \cdot 3^0 \cdot 5^1 & 2^2 \cdot 3^0 \cdot 5^1 \\ 2^0 \cdot 3^1 \cdot 5^0 & 2^1 \cdot 3^1 \cdot 5^0 & 2^2 \cdot 3^1 \cdot 5^0 \\ 2^0 \cdot 3^1 \cdot 5^1 & 2^1 \cdot 3^1 \cdot 5^1 & 2^2 \cdot 3^1 \cdot 5^1 \end{array}$$

This systematic listing makes it clear that there are three sets of combinations, beginning with $2^0$, $2^1$, or $2^2$. Within each of these sets there are two exponents of 3 which must be paired with the two exponents of 5, to give $2 \cdot 2 = 4$ combinations in each set. The total number of different divisors is then $3 \cdot 2 \cdot 2 = (2+1)(1+1)(1+1) = 12$. Sometimes a "tree diagram" such as the one shown in Fig. 1–1 below is used to illustrate all of the possible combinations.

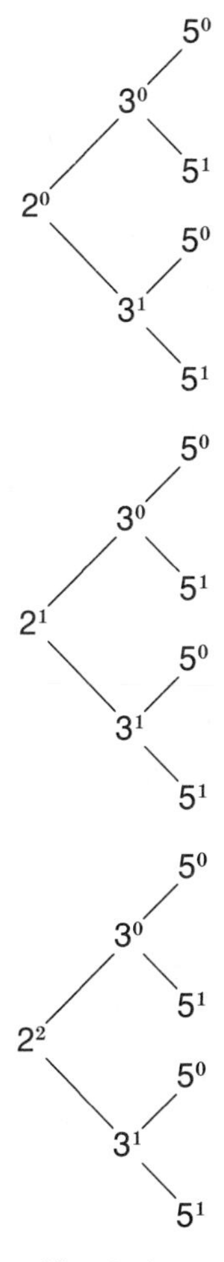


**Fig. 1–1**

***Some of the divisors of 60 are 1, 6, and 12. Find these divisors in the "tree diagram" above.***

In the following figure, the shaded areas from top to bottom contain the factorization of 1, 6, and 12, respectively.

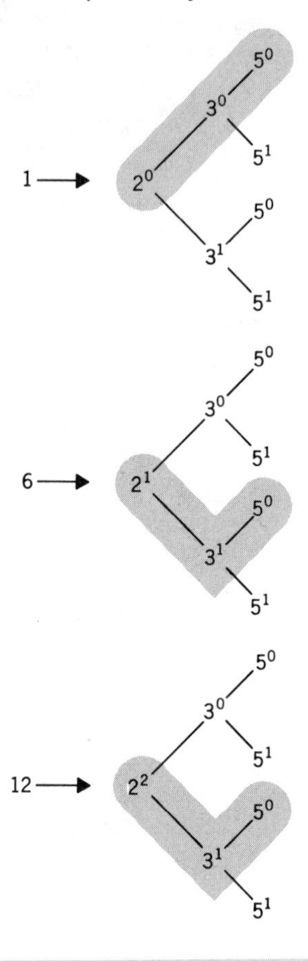


***List all the divisors of 90 in a "tree diagram."***

Now that the idea behind the proof has been illustrated, let's formulate the actual proof.

*Proof of Theorem:* From the lemma we know that every divisor not equal to 1 of $N$ is of the form $q_1^{b_1} \cdot q_2^{b_2} \cdot \ldots \cdot q_m^{b_m}$ where every $q$ is a prime $p$ in the prime factorization of $N$ and every $b$ is some integer between 1 and the corresponding $a$ of $p$ inclusive. If any $p$ in $N$ does not appear in the prime factorization of the divisor, we can put it in with an exponent of 0. The divisor 1 can also be written this way using all zero exponents. It is also true that any number formed the way just explained is in fact a divisor of $N$. (Can you explain why?) Thus we can list *all* of the divisors of $N$ by this method, that is by using all possible combinations of the exponents. The first exponent could be 0, 1, 2, ... up to $a_1$ inclusive, the second could be 0, 1, 2, ... up to $a_2$ inclusive, and so on until the last exponent which could be 0, 1, 2, ... up to $a_n$ inclusive. The total number of combinations of exponents is then $(a_1 + 1)(a_2 + 1) \ldots (a_n + 1)$ and this product equals $d(N)$. ■

## Exercises

**1.** Complete the proof of the Lemma which was started in this section.

**2.** What is the prime factorization of 180? List all the numbers you get by taking all possible exponents between 0 and each exponent in this factorization. Are all of these numbers divisors of 180? Are any divisors of 180 missing? How many divisors did you find this way? Does this number agree with that given by the formula for $d(N)$?

**3.** Find $d(2^{11} - 1)$, $d(468)$, $d(210)$, $d(360)$, and $d(720)$.

**4.** Find the smallest numbers with exactly 10, 18, and 24 divisors, respectively.

***5.** Prove that if $d(N)$ is odd, then $N$ is a perfect square. (A natural number $N$ is a perfect square if there is a natural number $M$ such that $N = M^2$.)

***6.** Prove that if $d(N)$ is even, then $N$ is not a perfect square.

***7.** Find $m(N)$, the product of all the divisors of $N$, for $1 \leq N \leq 25$; for example, $m(6) = 1 \cdot 2 \cdot 3 \cdot 6 = 36$, and $m(9) = 1 \cdot 3 \cdot 9 = 27$. Make a conjecture about the relationship between $m(N)$ and $N$. Prove or disprove your conjecture.

### 5. *d(N)* IS MULTIPLICATIVE

Some people think that mathematicians are people who like to calculate. The opposite is closer to the truth. A real mathematician dislikes long calculations and will work very hard to find a way to avoid them. So in finding a formula for $d(N)$ we have just accomplished some real mathematics. We have found a formula which expresses the *relationship* between $N$ and $d(N)$. Mathematicians dislike calculations, but do like to study relationships. Having found a relationship they seldom stop there; they look a little further for more relationships.

* Exercises preceded by an asterisk are more difficult than the rest.

In this spirit we might ask now whether the $d(N)$'s we have found are somehow related to each other.

***Look at d(3), d(5), and d(15). Are you able to find a relationship on these three numbers? Look at d(3), d(25), and d(75). Do you notice a similar relationship holding for these numbers?***

We should notice that $d(3) = 2$, $d(5) = 2$, $d(15) = 4$, and that $3 \cdot 5 = 15$ while $2 \cdot 2 = 4$. In the second case, $d(3) = 2$, $d(25) = 3$, $d(75) = 6$ and $3 \cdot 25 = 75$ while $2 \cdot 3 = 6$.

***Formulate a conjecture about the relationship between the d(N)'s.***

CONJECTURE: $d(N) \cdot d(M) = d(N \cdot M)$.

Again, this is only a conjecture; it may be true or false. We must either prove it, or produce a counterexample.

It doesn't take long to find a counterexample: $d(3) = 2$, $d(6) = 4$, $d(18) = 6$ and $3 \cdot 6 = 18$ while $2 \cdot 4 \neq 6$. We write $d(3) \cdot d(6) \neq d(18)$.

Our conjecture is false, however, this does not mean we should discard it entirely. It seems that $d(N) \cdot d(M) = d(N \cdot M)$ is true for many cases; perhaps we can modify the conjecture so that the new conjecture will be true.

Let's look at several cases in which the conjecture holds.

***Find six examples for which the conjecture does hold.***

We might list:

$$d(2) \cdot d(3) = (1+1)(1+1) = d(6),$$
$$d(5) \cdot d(7) = (1+1)(1+1) = d(35),$$
$$d(4) \cdot d(3) = (2+1)(1+1) = d(12),$$
$$d(2) \cdot d(9) = (1+1)(2+1) = d(18),$$
$$d(6) \cdot d(5) = (1+1)(1+1)(1+1) = d(30),$$
$$d(2) \cdot d(25) = (1+1)(2+1) = d(50).$$

***Now find six examples for which the conjecture does not hold.***

Some examples are:

$$d(2) \cdot d(2) = (1+1)(1+1) \neq (2+1) = d(4),$$
$$d(2) \cdot d(6) = (1+1)(1+1)(1+1) \neq (2+1)(1+1) = d(12),$$
$$d(3) \cdot d(3) = (1+1)(1+1) \neq (2+1) = d(9),$$
$$d(3) \cdot d(12) = (1+1)(2+1)(1+1) \neq (2+1)(2+1) = d(36),$$
$$d(5) \cdot d(10) = (1+1)(1+1)(1+1) \neq (1+1)(2+1) = d(50),$$
$$d(2) \cdot d(14) = (1+1)(1+1)(1+1) \neq (2+1)(1+1) = d(28).$$

There must be something different about these two sets of examples. Why should $d(N) \cdot d(M) = d(N \cdot M)$ be true for the first set and not true for the second?

***Can you see some difference between these two sets?***

In the first set we find the pairs 2 and 3, 5 and 7, 4 and 3, 2 and 9, 6 and 5, and 2 and 25. In each case there is no common divisor of both numbers except the number 1. In the second set we find the pairs 2 and 2, 2 and 6, 3 and 3, 3 and 12, 5 and 10, and 2 and 14. For each pair there is a common divisor other than 1; namely, 2, 2, 3, 3, 5, and 2, respectively. It seems that our conjecture might be true if we restrict it to two numbers which have no common divisor other than 1. (Two natural numbers which have no common divisor other than 1 are called *relatively prime.)* This modified conjecture is correct, so we'll call it a theorem.

THEOREM: If $N$ and $M$ have no common divisors other than 1, then $d(N) \cdot d(M) = d(N \cdot M)$.

*Proof:* Let $N$ have prime factorization $p_1^{a_1} \cdot p_2^{a_2} \cdot \ldots \cdot p_n^{a_n}$. Let $M$ have prime factorization $q_1^{b_1} \cdot q_2^{b_2} \cdot \ldots \cdot q_m^{b_m}$. Since $N$ and $M$ have no common factors, a prime factorization of $N \cdot M$ is $p_1^{a_1} \cdot p_2^{a_2} \cdot \ldots \cdot p_n^{a} \cdot q_1^{b_1} \cdot q_2^{b_2} \cdot \ldots \cdot q_m^{b_m}$. Then $d(N) \cdot d(M) = (a_1+1)(a_2+1) \ldots (a_n+1)(b_1+1)(b_2+1) \ldots (b_m+1)$, which is the value of $d(N \cdot M)$. ■

Since in many cases we can now *multiply* previously calculated $d(N)$'s to find a new one, let's say that $d(N)$ is *multiplicative.** For example,

* Because of the restriction which requires that the factors be relatively prime, some mathematicians use the term "almost multiplicative."

suppose we want to find $d(150)$, we can take advantage of the fact that $150 = 3 \cdot 50$. Now 3 and 50 have no common divisor other than 1, and we already have found $d(3)$ and $d(50)$, so

$$\begin{aligned} d(150) &= d(3 \cdot 50) \\ &= d(3) \cdot d(50) \\ &= 2 \cdot 6 \\ &= 12. \end{aligned}$$

## Exercises

**1.** Find $d(21)$, $d(42)$, $d(45)$, $d(20)$, and $d(36)$ using the fact that $d(N)$ is multiplicative, that is, $d(N \cdot M) = d(N) \cdot d(M)$ if $N$ and $M$ have no common divisors other than 1.

**2.** Is $m(N)$ defined in Exercise 7 of the last section multiplicative, that is, if $N$ and $M$ have no common divisors other than 1, does $m(N) \cdot m(M) = m(N \cdot M)$? Why, or why not?

**3.** Let $s(N)$ be the sum of all of the divisors of $N$. For example, $s(1) = 1$, $s(2) = 1 + 2 = 3$, $s(6) = 1 + 2 + 3 + 6 = 12$. Find $s(N)$ for $1 \leq N \leq 25$. Does it seem that $s(N)$ is multiplicative?

***4.** Find a formula for calculating $s(N)$ as just defined. Prove that $s(N)$ is multiplicative.

**5.** Let $a(N)$ be the sum of the digits in $N$. For example, if $N = 25$, then $a(25) = 2 + 5 = 7$. Is $a(N)$ multiplicative? Why, or why not?

**6.** Let $t(N)$ be the sum of all $d(M)$ such that $M$ is a divisor of $N$. For example, $t(12) = d(1) + d(2) + d(3) + d(4) + d(6) + d(12) = 1 + 2 + 2 + 3 + 4 + 6 = 18$. Calculate $t(N)$ for $1 \leq N \leq 25$. Does $t(N)$ seem to be multiplicative?

### 6. AN ORDERING FOR THE DIVISORS OF A NUMBER

We have just studied two different relationships—one between $N$ and $d(N)$, the other concerns certain $d(N)$, $d(M)$, and $d(N \cdot M)$. These are only two of the many relationships that are about divisors of numbers. (In the exercises, $m(N)$, $s(N)$, and $t(N)$ were three more notions related to $N$ and concerned with the divisors of $N$.)

Let's look now at a single number and all of its divisors and see if we can find still another relationship.

***List all of the divisors of 12.***

The divisors of 12 are 1, 2, 3, 4, 6, and 12. We probably were not conscious of it, but we used a relationship between these divisors to write them down in an orderly way. We listed the divisors in order according to size. When we say that one number is *less than* another number we are talking about a relationship between two numbers. For some numbers $A$ and $B$ in that order, we can say $A$ is less than $B$, and write $A < B$. For example, $2 < 4$. For other numbers in a given order, say 6 and 3, we cannot say that the relation $<$ holds between the numbers (in that order).

Of course, the relation $<$ applies to more than just the divisors of 12. For any two natural numbers $A$ and $B$, either $A < B$, or this is not so. Since the relation $<$ applies to all natural numbers, we do not want to study it here. We would like to find a relation which applies specifically to the divisors of a number.

***Look at the divisors of 12 and see if you can find a relation which holds for some pairs of divisors in a certain order and does not hold for other pairs.***

We should notice that some divisors are divisors of other divisors, and some are not. For example, 1 divides 2, 1 divides 3, 2 divides 4, 2 divides 6, 6 divides 12; however, 2 does not divide 3, 4 does not divide 6, and 4 does not divide 2, and so on. If $A$ divides $B$, we write $A|B$. If $A$ does not divide $B$, we write $A \nmid B$.

We could keep track of which numbers divide others by the following diagram in Fig. 1–2.

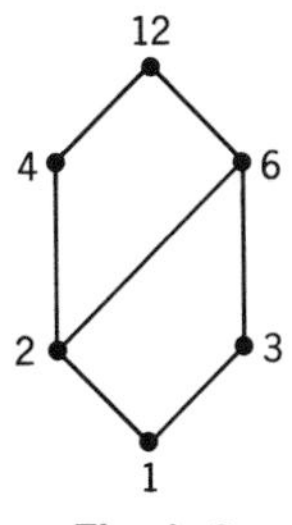


**Fig. 1–2.**

If one number is below another and connected to it by an ascending line, then the lower number is a divisor of the upper number. The diagram shows that 1 divides all the other numbers, and that $2|4$, $2|6$, $2|12$, $3|6$, $3|12$, $4|12$, and $6|12$. We call this diagram a *lattice*. The lattice also shows us that $2 \nmid 3$ and $4 \nmid 6$. Of course, only a number below another can divide it, so $4 \nmid 2$, $6 \nmid 3$, $6 \nmid 2$, and so on.

Another diagram which represents the same relationship is:

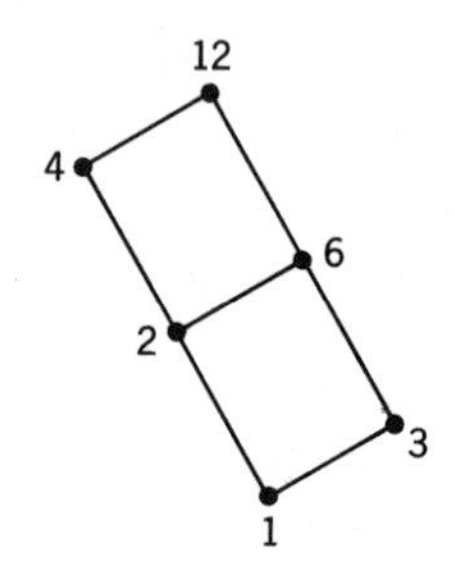


There is no essential difference between these two diagrams because they describe exactly the same relationship. For our purposes we will make no distinction between these two diagrams.

---

***Draw the lattices for the divisors of 7, of 16, of 15, and of 24.***

---

We should draw diagrams similar to the following:

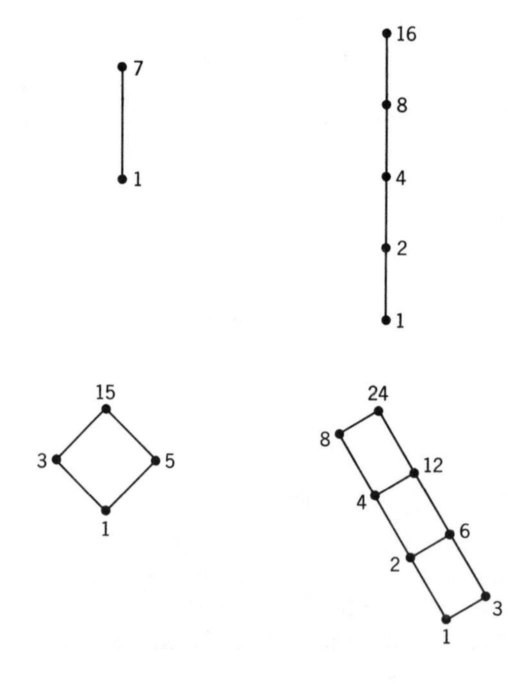

These examples help us to see that lattices are a generalization of the way the relation $<$ orders numbers. The lattices for the divisors of 7 and 16 look like the numbers have been ordered by $<$. The lattices for the divisors of 15 and 24 are not this simple, but they certainly do arrange or *order* the divisors in a definite way.

It would be more correct to say that | is a generalization of $\leq$. For two numbers $A$ and $B$, we write $A \leq B$ if $A$ is less than $B$ or if $A$ equals $B$. Thus, $2 \leq 3$ and $2 \leq 2$. Similarly, when we write $A|B$, $A$ may be a proper divisor of $B$ ($A|B$ and $A < B$), *or* $A$ may be equal to $B$. In a lattice this means that one number divides another if the first is below the second and is connected to it by an ascending line *and* every number is a divisor of itself.

## Exercises

1. Draw the lattice diagrams for the divisors of 11, 8, 18, 27, 30, 36, and 100.
2. Look at the lattice of divisors of 24. What is the largest number that divides 1 and 2? 2 and 3? 4 and 6? 8 and 12? 8 and 4? 3 and 6? 4 and 24? 4 and 4? Each of these numbers is called the *greatest common divisor* of the two given numbers. The greatest common divisor of $A$ and $B$ is symbolized by gcd($A$, $B$). Describe the position of the gcd($A$, $B$) in the lattice relative to $A$ and $B$.
3. Look at the lattice of divisors of 24. A multiple of a number $N$ is some number $K = N \cdot T$ where $T$ is any natural number. For example, 12 is a multiple of 3 because $12 = 3 \cdot 4$. What is the smallest number which is a multiple of both 2 and 3? 1 and 2? 4 and 6? 8 and 4? 3 and 6? 4 and 24? 4 and 4? The smallest number which is a multiple of $A$ and $B$ is called the *least common multiple* of $A$ and $B$ and is written lcm($A$, $B$). Describe the position in the lattice of the lcm($A$, $B$) relative to $A$ and $B$.

## 7. WHAT HAVE WE LEARNED FROM DIVISORS?

We have just seen a little piece of number theory. We now know what is meant by divisors, $d(N)$, primes, composites, and unique factorization. More important, our study of divisors has shown us many facets of mathematics. Let's summarize them.

***Can you list some characteristics of mathematics which were illustrated by our study of divisors?***

We have found out that mathematicians sometimes do work with numbers, however, perhaps not in the way we expected. Mathematicians are not particularly interested in adding long columns of numbers or in doing a long division problem. *Mathematicians are interested in finding relationships between numbers.* This was illustrated by the relationship between $d(N)$ and $N$, the relationship among certain $d(N)$, $d(M)$, and $d(N \cdot M)$, and also by the ordering of the divisors of a number in a lattice by the relation $|$.

Our study also brought out that *mathematicians experiment.* The many computations and lists which we wrote out are good examples of what a mathematician does in order to discover relationships. He looks at many particular cases, gets an intuitive feeling for what is happening, and then *makes a conjecture.* In this respect he is very much like the experimental scientist. Most people are unaware of this aspect of a mathematician's work; these people oversimplify the situation and say that a mathematician uses only deductive reasoning. People get this impression because they usually see only the finished mathematical theorem and its proof; only deductive reasoning is permitted here. They do not witness, however, all of the preliminary work that led up to the formulation of the proof.

Sometimes the conjecture of a mathematician is not true. If this is the case he must show that it is false by means of a *counterexample.* If the conjecture is in the form *if A, then B,* a counterexample will be a specific case in which the *if* clause is true and the *then* clause is false.

False conjectures are not always useless. Often they can be modified to form true statements. A true mathematical statement which can be proved is called a *theorem.* (There are true mathematical statements which cannot be proved. See Section 6 of Chapter 4.) Various methods of proof are used for theorems. In this chapter we saw a *proof by contradiction* in the proof of the lemma. To prove *if A, then B* by contradiction, we assume that *not-B* is true, then reason from *A and not-B* until we reach a contradiction. This means that *A and not-B* is false, or that *if A, then B* must be true.

An example of a proof by contradiction outside of mathematics may help to clarify this. Suppose we wanted to prove that

> If it is snowing,
> then we need to wear coats outside.

We can prove this by assuming the negative of the *then* clause, *we do not need to wear coats outside,* and the *if* clause, *it is snowing.* Since we do not need coats it is reasonable to conclude that the temperature must be above 40°. Since it is snowing, it is also reasonable to conclude the temperature must be below 40°. The temperature cannot be above and below 40° at the same time, so we have reached a contradiction. Thus our assumption that we do not need coats cannot be true if it is snowing, so we do need to wear coats outside.

Another thing that we should notice is that in mathematics we *use symbols* to represent a general situation (as we did when we wrote out the prime factorization of any $N$). These symbols provide an accurate way to talk about general cases. The proof of a theorem always has a general character about it. Because of this highly specialized symbolic language, many people feel that mathematics is very difficult and far beyond their comprehension. A laundry list read in fluent Chinese could seem very eloquent to someone who didn't speak the language. Mathematics is not as difficult as people think, but there is a language barrier. Mathematicians know their language so well that they literally think in it, as the person who has mastered French will find himself thinking, and perhaps dreaming, in French.

Finally, in our work with lattices we saw that the mathematician tries to *generalize* previously known relationships. The lattice for $|$ is a generalization of the ordering of numbers by $\leq$. Many new ideas are introduced into mathematics by this process of generalization. We will see several more examples of generalization in later chapters.

## Exercises

**1.** Give counterexamples for the following false conjectures:

(a) If $N|M$, then $M|N$.

(b) If $N|M \cdot R$ and $N \nmid M$, then $N|R$.

(c) If $N \nmid M$ and $N \nmid R$, then $N \nmid M \cdot R$.

***2.** Experiment and find a modification (or modifications) of the conjecture in Exercise 1(c) which is true. Prove it.

**3.** Prove by contradiction that:

> If green is yellow and blue, brown is green and red, and purple is red and blue, then brown is not purple and red.

### 8. WHAT IS NUMBER THEORY?

Number theory is a branch of mathematics in which we study the properties of the integers, 0, 1, −1, 2, −2, . . . . Our study of divisors belongs to number theory.

Some typical questions that are raised in number theory are:

1. How many primes are there?
2. How many pairs of primes differ by 2?
3. Can every even integer greater than 2 be written as the sum of two primes?
4. What are the solutions in integers of $2x + 3y = 1$?
5. What are the solutions in integers of $x^2 + y^2 = z^2$?
6. Does $x^4 + y^4 = z^4$ have any solutions in integers other than $x = 0$, $y = 0$, and $z = 0$?

***Before we answer the questions see what you think about them. Talk to your friends about them.***

The first question about the number of primes was answered by Euclid in 300 BC. He proved that no matter how many primes are listed, it is always possible to find another prime that has not yet been listed. This proved that the number of primes is infinite.

The second question is as easy to understand as the first. There are some pairs of primes which differ by 2, for example, 11 and 13, 17 and 19, 29 and 31. How many such pairs are there? Although the question is easy to understand, no one to this date has answered it. This is called an unsolved problem in number theory. Number theorists feel that there are an infinite number of such pairs of primes, but no one has proved it.

In 1742 Christian Goldbach made a conjecture about Question 3. It seems that every even integer greater than 2 can be written as the sum of two primes, for example $8 = 5 + 3$, $16 = 5 + 11$ or $16 = 13 + 3$, $38 = 19 + 19$, and $54 = 13 + 41$. Goldbach conjectured that the answer to Question 3 is yes. Like the second question, this one also has not yet been settled. It, too, is an unsolved problem in number theory.

We shouldn't begin to think that number theory consists of only unsolved problems. The fourth question has been answered: $x = -1$, $y = 1$ is one solution, $x = -4$, $y = 3$ is another solution, and we can find all possible solutions by letting $x = -1 - 3t$ and $y = 1 + 2t$ where $t$ is any integer. In number theory we find a way to arrive at this solution. Question 4 has been completely answered.

Question 5 has been answered also. One solution is $x = 3$, $y = 4$, and $z = 5$. There is an infinity of solutions and number theorists know how to find them. Another solution is $x = 5$, $y = 12$, and $z = 13$. (Notice that if we were studying geometry these solutions would tell us the dimensions of the right triangles with integral sides.)

The last question is closely related to the previous one. The equation $x^4 + y^4 = z^4$ has no solution in integers other than $x = 0$, $y = 0$, and $z = 0$. It is also related to an unsolved problem. In the 17th century, Fermat conjectured that $x^n + y^n = z^n$ has no solution other than $x = 0$, $y = 0$, and $z = 0$, if $n$ is greater than 2. Fermat may have had a proof of this conjecture, but if he did he never wrote it down. No one else to this day has been able to prove it.

One thing which is illustrated by these questions is that most of the results of number theory are easy to understand since they are about integers. However, the proofs of many of the theorems are very complicated and there are many conjectures which still have not been proved.

Number theorists sometimes pride themselves on studying the branch of mathematics which is so pure that it has no practical uses. (See Hardy's *A Mathematician's Apology.*) They study it just for the beauty of the subject itself. It is not completely true, however, that number theory has no applications.

One extremely practical use of number theory is in computers. Centuries before the invention of the computer, number theorists studied numeration systems other than our usual decimal system. In the decimal system we use ten symbols 0, 1, 2, 3, 4, 5, 6, 7, 8, and 9 and the places represent powers of 10; for example, 256 means 6 *ones* (or $10^0$s), 5 *tens* (or $10^1$s), and 2 *hundreds* (or $10^2$s). A numeration system need not be based on 10; it could be based on 2, 5, 6, 8, 12, or any other natural number greater than 1. In a numeration system based on 2 we use two symbols, 0 and 1, and the places represent the powers of 2. For example, 1011 means 1 *one* (or $2^0$), 1 *two* (or $2^1$), 0 *fours* (or $2^2$s), and 1 *eight* (or $2^3$). When computers were invented this binary (two symbol) numeration system was exactly what best fitted their construction. (See Chapter 9 for a more complete explanation.)

If binary arithmetic had not been studied already, it probably would have been investigated when computers were beginning to be invented. Fortunately, it had already been developed because mathematicians were simply interested in studying a numeration system for its own sake. Certainly the first man who did binary arithmetic never dreamt that he was making a contribution to putting a man on the moon. This example should be a warning to us: we should be cautious about saying that some part of mathematics is useless.

Another direct use of number theory has been in coding. Besides these direct applications, number theory is frequently used in other branches of mathematics. Modern algebra in particular draws heavily on number theory. Modern algebra in turn is now a common tool of the chemist and physicist. Some of the specific applications of modern algebra will be mentioned at the end of Chapter 2.

## Exercises

**1.** Look up the lives of Pierre de Fermat, Leonhard Euler, Carl Friedrich Gauss, Joseph Louis Lagrange, or Adrien Marie Legendre. Notice in particular the contributions each made to number theory.

**2.** Read one of the following:

Dunnington, Guy Waldo, *Carl Friedrich Gauss, Titan of Science,* Exposition Press, New York, 1955.

Hardy, G. H., *A Mathematician's Apology,* Cambridge University Press, Cambridge, 1941. (An excerpt is in Volume 4 of *The World of Mathematics,* J. R. Newman, Simon and Schuster, New York, 1956, pp. 2024–2038.)

*Herwitz, P. S., "The Theory of Numbers," *Scientific American,* Vol. 185, No. 1, July, 1951, pp. 52–55.

*Newman, James R., "Srinivasa Ramanujan," *Scientific American,* Vol. 178, No. 6, June, 1948, pp. 54–57.

Ogilvy, C. Stanley, and John T. Anderson, *Excursions in Number Theory,* Oxford University Press, New York, 1966.

Schaaf, William L., *Carl Friedrich Gauss,* Watts, New York, 1964.

The article on Theory of Numbers in a recent edition (1966 to the present) of the *Encyclopaedia Britannica.*

## Test

**1.** What is a prime number?

**2.** Find the prime factorization of: 5, 8, 24, 30, 36.

**3.** What does the expression "$D$ divides $N$" mean?

**4.** Use Problem 2 and the formula for $d(N)$ to find: $d(5)$, $d(8)$, $d(24)$, $d(30)$, $d(36)$.

**5.** Use the lemma proved in this chapter to explain why 8 cannot be a divisor of 100.

**6.** Use the multiplicative property of $d(N)$ to find $d(15 \cdot 16)$.

**7.** Find the smallest number with exactly 5 divisors.

**8.** Draw the lattice of divisors of 42.

**9.** Give a counterexample for: If $N|M$, $N|R$, and $R|M$, then $N|(M/R)$. ($M/R$ means the quotient, $M$ divided by $R$.)

**10.** Report on your outside reading. You were asked to find out about the lives of Fermat, Euler, Gauss, Lagrange, or Legendre; or you should have found some facts additional to this text about the nature of number theory. Which of these did you do? Summarize what you found.

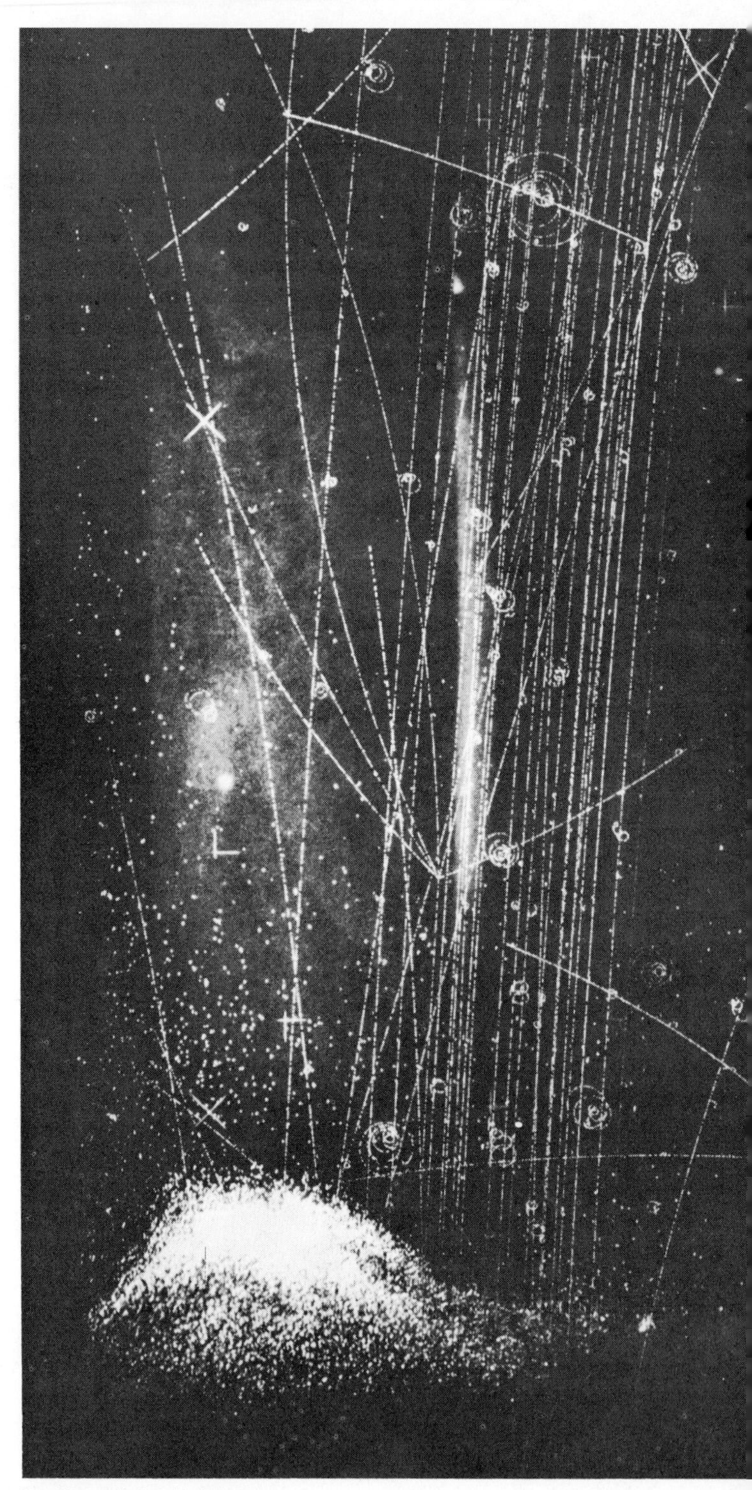

Brookhaven National Laboratory

2

# MODERN ALGEBRA

## Multiplying Without Numbers

Modern algebra, especially group theory, plays a large role in the study of the elementary particles whose traces are recorded in this bubble-chamber photograph.

In case some of us feel that mathematics is mostly about numbers, let's look at a branch of mathematics in which numbers are not the main interest. We will look at a topic in modern algebra—*groups.*

## 1. AN EXAMPLE OF A GROUP

Let's imagine that we have three books labeled $X$, $Y$, $Z$ arranged in some order from left to right on a shelf. We would like to rearrange the books and we wonder what different rearrangements are possible.

For example, if we label the positions $a$, $b$, $c$, and the books are originally arranged so that $X$ is at $a$, $Y$ is at $b$, and $Z$ is at $c$

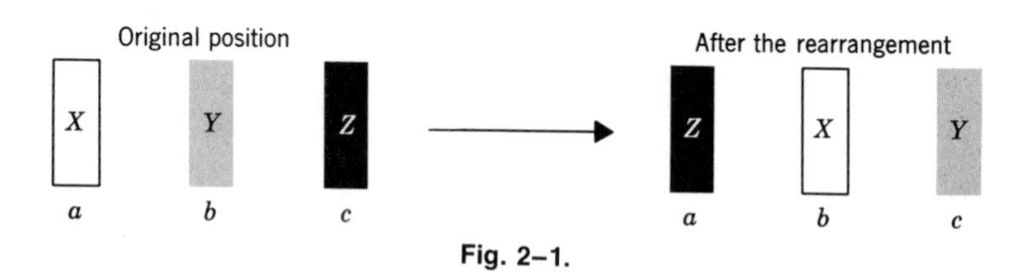


**Fig. 2–1.**

we could move the book $X$ which is at $a$ to the $b$ position, move the book $Y$ at $b$ to the $c$ position, and the book $Z$ at $c$ to the $a$ position (Fig. 2–1). We can indicate this by $a \rightarrow b$, $b \rightarrow c$, $c \rightarrow a$, or by the abbreviation

$$\begin{array}{ccc} a & b & c \\ \downarrow & \downarrow & \downarrow \\ b & c & a, \end{array}$$

or finally $\begin{pmatrix} a & b & c \\ b & c & a \end{pmatrix}$. WE READ EACH COLUMN FROM TOP TO BOTTOM AS FOLLOWS: "THE BOOK AT $a$ MOVES TO $b$, THE BOOK AT $b$ MOVES TO $c$, AND THE BOOK AT $c$ MOVES TO $a$."

***Draw pictures as above to illustrate all possible rearrangements of three books. Represent these rearrangements by symbols.***

Besides $\begin{pmatrix} a & b & c \\ b & c & a \end{pmatrix}$ we could rearrange the books according to $\begin{pmatrix} a & b & c \\ c & a & b \end{pmatrix}$ and the diagram would be:

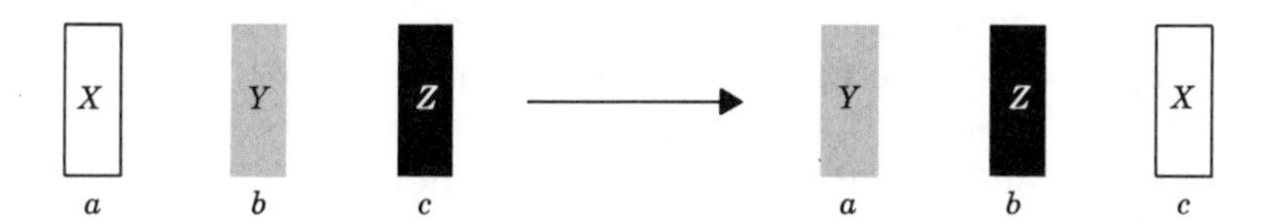


The following rearrangements which leave one book fixed are also possible: $\begin{pmatrix} a & b & c \\ a & c & b \end{pmatrix}$, $\begin{pmatrix} a & b & c \\ c & b & a \end{pmatrix}$, and $\begin{pmatrix} a & b & c \\ b & a & c \end{pmatrix}$. We might also leave all the books as they are; this is represented by $\begin{pmatrix} a & b & c \\ a & b & c \end{pmatrix}$. If we include this last case, there are six possible rearrangements of the books. Let's give these rearrangements single letter names so that we can talk about them more easily.

Let

$$A = \begin{pmatrix} a & b & c \\ a & b & c \end{pmatrix} \qquad B = \begin{pmatrix} a & b & c \\ b & c & a \end{pmatrix} \qquad C = \begin{pmatrix} a & b & c \\ c & a & b \end{pmatrix}$$

$$D = \begin{pmatrix} a & b & c \\ a & c & b \end{pmatrix} \qquad E = \begin{pmatrix} a & b & c \\ c & b & a \end{pmatrix} \qquad F = \begin{pmatrix} a & b & c \\ b & a & c \end{pmatrix} .$$

Since there are six rearrangements we might ask what would happen if we follow one rearrangement by another rearrangement. For example, we might rearrange the books according to $B$ and then rearrange them

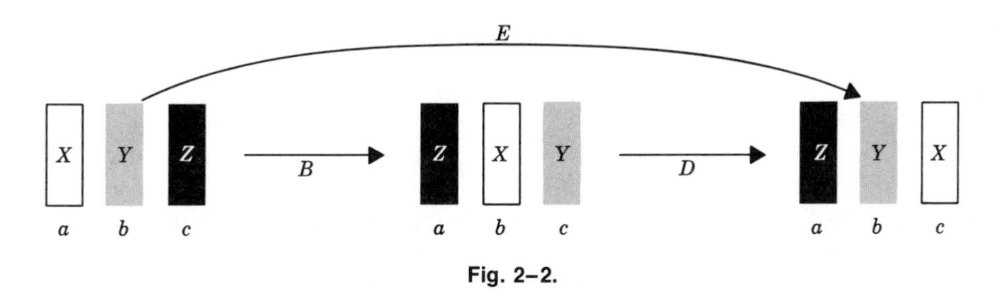


**Fig. 2–2.**

again by $D$ as shown in Fig. 2–2. Rearrangement $B$ would take the book at $a$ to position $b$ and then rearrangement $D$ would take it to position $c$. Similarly, the book at $b$ would go to position $c$, then back to $b$, and the book at $c$ would go to $a$, then stay at $a$. The total effect would be $\begin{pmatrix} a & b & c \\ c & b & a \end{pmatrix}$.

It should not surprise us that this is $E$, one of the rearrangements on our list. We would expect that two rearrangements performed in succession would be a rearrangement. Since we have listed all possible rearrangements, it should be on our list.

We could represent these two rearrangements performed in succession by $BD = E$, in which we mean $B$ is performed first, then $D$, and the total effect is $E$. We seem to have found something that resembles the operation of multiplication. One number times another number gives us a third number. One rearrangement followed by a second rearrangement gives us a third rearrangement. We will call $BD$ the *"product"* of $B$ and $D$.

---

***Find* BE, DE, BA, AA, CC, FE. *Draw pictures as above to help you for the first few "products," then try to find the "product" using only the symbols for the rearrangement.***

---

Since $B = \begin{pmatrix} a & b & c \\ b & c & a \end{pmatrix}$ and $E = \begin{pmatrix} a & b & c \\ c & b & a \end{pmatrix}$, then $BE$ will move the book in $a$ to position $b$, then keep it at $b$, the book in $b$ moves to $c$ then to $a$, and the book in $c$ to $a$ then back to $c$. This means that $BE = \begin{pmatrix} a & b & c \\ b & a & c \end{pmatrix} = F$. Similarly we should find that $DE = C$, $BA = B$, $AA = A$, $CC = B$, and $FE = B$.

We could keep track of these "products" by means of a table:

Second Rearrangement

| | | $A$ | $B$ | $C$ | $D$ | $E$ | $F$ |
|---|---|---|---|---|---|---|---|
| | $A$ | $A$ | | | | | |
| | $B$ | $B$ | | | $E$ | $F$ | |
| First | $C$ | | | $B$ | | | |
| Rearrangement | $D$ | | | | | $C$ | |
| | $E$ | | | | | | |
| | $F$ | | | | | $B$ | |

The column at the left will be called the *guide column* and indicates the first rearrangement. The row at the top will be called the *guide row* and indicates the second rearrangement. The entry in the body of the table is the "product."

---

***Complete the table of "products" of rearrangements of three books.***

---

We should get the following table:

| | A | B | C | D | E | F |
|---|---|---|---|---|---|---|
| A | A | B | C | D | E | F |
| B | B | C | A | E | F | D |
| C | C | A | B | F | D | E |
| D | D | F | E | A | C | B |
| E | E | D | F | B | A | C |
| F | F | E | D | C | B | A |

This set of six rearrangements with their "products" as indicated in this table is an example of a *group.* In this chapter we will study some of the properties of groups. The study of the properties of groups is a part of modern algebra. In a sense, in a group we can multiply without numbers. In our example instead of multiplying numbers we found the "products" of rearrangements.

## Exercises

**1.** List all the rearrangements of two books. Find the table which indicates the "products" of these rearrangements.

**2.** List all the rearrangements of four books. (There are twenty-four in all.)

**3.** Name the twenty-four rearrangements of four books as in the answer to Exercise 2 in the back of this book. Make a table which indicates the "products" of these twenty-four rearrangements. (Get together with some of your friends to do this problem and divide up the work.)

## 2. GROUPS

In Section 1 and the exercises that followed, we found three examples of groups:

The Rearrangements
of Two Books

| | A | B |
|---|---|---|
| A | A | B |
| B | B | A |

GROUP 1

Group 1 was found in Exercise 1, Section 1.

The Rearrangements
of Three Books

| | $A$ | $B$ | $C$ | $D$ | $E$ | $F$ |
|---|---|---|---|---|---|---|
| $A$ | $A$ | $B$ | $C$ | $D$ | $E$ | $F$ |
| $B$ | $B$ | $C$ | $A$ | $E$ | $F$ | $D$ |
| $C$ | $C$ | $A$ | $B$ | $F$ | $D$ | $E$ |
| $D$ | $D$ | $F$ | $E$ | $A$ | $C$ | $B$ |
| $E$ | $E$ | $D$ | $F$ | $B$ | $A$ | $C$ |
| $F$ | $F$ | $E$ | $D$ | $C$ | $B$ | $A$ |

GROUP 2

The table for Group 2 was developed in Section 1.

Group 3 was computed in Exercise 3, Section 1. The table is given on page 41.

Let's forget now about the books that led to these tables and just look at the tables themselves. Let's see if we can find some properties common to the "products" indicated by all three tables. Since each table specifies the "products" of a group some of the common properties can be used to describe the nature of a group.

***Stop and see if you can find any properties that these three tables have in common.***

We might notice the following common properties:

1. Each table deals with a certain set of objects which is the same as the set in the guide column and the guide row.
2. Every pair of objects in the set has exactly one "product."
3. The "products" are members of the set.
4. Every member of the set appears exactly once in each row and each column.
5. There is a column which looks exactly like the guide column.
6. There is a row which looks exactly like the guide row.
7. The "products" have the associative property, for all $X, Y, Z$ in the set, $(XY)Z = X(YZ)$.

The Rearrangements of Four Books

| | A | B | C | D | E | F | G | H | I | J | K | L | M | N | O | P | Q | R | S | T | U | V | W | X |
|---|---|---|---|---|---|---|---|---|---|---|---|---|---|---|---|---|---|---|---|---|---|---|---|---|
| A | A | B | C | D | E | F | G | H | I | J | K | L | M | N | O | P | Q | R | S | T | U | V | W | X |
| B | B | C | A | L | F | J | I | D | K | E | G | H | P | M | S | N | X | U | W | R | T | Q | O | V |
| C | C | A | B | H | J | E | K | L | G | F | I | D | N | P | W | M | V | T | O | U | R | X | S | Q |
| D | D | J | G | E | A | H | L | K | B | I | F | C | Q | U | M | W | O | S | V | P | X | R | T | N |
| E | E | I | L | A | D | K | C | F | J | B | H | G | O | X | Q | T | M | V | R | W | N | S | P | U |
| F | F | K | H | B | L | G | A | J | E | C | D | I | S | V | X | R | P | Q | U | O | M | W | N | T |
| G | G | D | J | K | I | A | F | C | L | H | B | E | U | W | T | Q | R | P | M | X | S | N | V | O |
| H | H | F | K | J | C | L | D | I | A | G | E | B | V | R | N | S | W | O | X | M | Q | T | U | P |
| I | I | L | E | G | K | B | J | A | H | D | C | F | T | O | R | X | U | N | P | V | W | M | Q | S |
| J | J | G | D | C | H | I | B | E | F | A | L | K | W | Q | V | U | N | X | T | S | P | O | M | R |
| K | K | H | F | I | G | C | E | B | D | L | A | J | R | S | U | V | T | M | N | Q | O | P | X | W |
| L | L | E | I | F | B | D | H | G | C | K | J | A | X | T | P | O | S | W | Q | N | V | U | R | M |
| M | M | N | P | O | Q | V | T | S | U | X | R | W | A | B | D | C | E | K | H | G | I | F | L | J |
| N | N | P | M | W | V | X | U | O | R | Q | T | S | C | A | H | B | J | I | L | K | G | E | D | F |
| O | O | X | T | Q | M | S | W | R | N | U | V | P | E | I | A | L | D | H | F | C | J | K | G | B |
| P | P | M | N | S | X | Q | R | W | T | V | U | O | B | C | L | A | F | G | D | I | K | J | H | E |
| Q | Q | U | W | M | O | R | P | V | X | N | S | T | D | J | E | G | A | F | K | L | B | H | C | I |
| R | R | S | V | U | T | P | Q | N | O | W | M | X | K | H | I | F | G | A | B | E | D | C | J | L |
| S | S | V | R | X | P | W | O | U | M | T | Q | N | F | K | B | H | L | D | J | A | E | G | I | C |
| T | T | O | X | R | U | M | V | P | W | S | N | Q | I | L | G | E | K | C | A | J | H | B | F | D |
| U | U | W | Q | T | R | N | X | M | S | O | P | V | G | D | K | J | I | B | C | F | L | A | E | H |
| V | V | R | S | N | W | T | M | X | Q | P | O | U | H | F | J | K | C | E | I | D | A | L | B | G |
| W | W | Q | U | V | N | O | S | T | P | R | X | M | J | G | C | D | H | L | E | B | F | I | K | A |
| X | X | T | O | P | S | U | N | Q | V | M | W | R | L | E | F | I | B | J | G | H | C | D | A | K |

GROUP 3

The first six properties can be seen easily in the tables. The seventh property is not immediately apparent. We are all familiar with the associative property for ordinary multiplication. If we want to multiply $2 \cdot 3 \cdot 5$ we know that we can do the problem two ways. We indicate this with parentheses and say $(2 \cdot 3) \cdot 5 = 2 \cdot (3 \cdot 5)$. All of the "products" defined by the tables have a similar associative property, $(XY)Z = X(YZ)$. The notation $(XY)Z$ means first find the "product" $V = XY$, and then the "product" $VZ$. The notation $X(YZ)$ means first find the "product" $U = YZ$, and then the "product" $XU$.

---

***Using Group 1, compute the following:* (AA)A *and* A(AA), (AA)B *and* A(AB), (AB)A *and* A(BA), (BA)A *and* B(AA), (AB)B *and* A(BB), (BA)B *and* B(AB), (BB)A *and* B(BA), *and* (BB)B *and* B(BB).**

---

In all possible cases the pairs of "products" are equal, so Group 1 has the associative property. Groups 2 and 3 also have this property, however, it would take a long time to check all the cases!

---

***How many cases would we have to check to show that Group 2 is associative? Group 3?***

---

Since we would have to check all possible "products" of three factors and we have six objects in Group 2, there are $6^3 = 216$ different cases to check. For Group 3 there would be $(24)^3$ cases! We will not check all of these, and, of course, mathematicians usually do not do so either. If possible, they find a short general way to prove that the "products" given by a table have this property without going through all those computations.

Even though we are not going to check every case, let's check a few so that at least we will *feel* that the tables are associative.

---

***In Group 2, show that* (AB)C = A(BC), (DE)F = D(EF), (AC)D = A(CD), *and* (BD)F = B(DF). *In Group 3, show that* (FJ)M = F(JM), (CE)L = C(EL), (RM)T = R(MT), *and* (XH)Q = X(HQ).**

---

We will agree that a *group* is a set of objects with "products" specified by a table having Properties 1 to 7. The set of objects in the group is the same as the set of objects in the guide row and column. Since we are using tables to determine the "products" in our groups, this implies that there are only a finite number of elements in the groups we are considering. We would be more correct if we would say that a *finite group* is a set of objects with "products" specified by a table with Properties 1 to 7. Unless otherwise specified, the word "group" in this chapter means "finite group."

Infinite groups also exist, for example, the integers (. . . , −2, −1, 0, 1, 2, . . .) with the operation of ordinary addition playing the role of "product." Although the operation is not specified by a table, addition of the integers does have the equivalent of all seven properties (See Exercises 6 and 7). We should not be disturbed by the fact that we are suddenly talking about an operation of addition, when up to this point we spoke of "products." "Product" simply means an operation. The operation may be like multiplication, like addition, or even something totally different than these familiar operations. From this point on in this chapter we will omit the quotation marks around "product." Product will mean the operation of the group.

We might have noticed that the three tables of rearrangements had some other common properties. These common properties may or may not be group properties. The fact that all three groups have an even number of elements is not a group property. We will soon see groups with an odd number of elements. The fact that the row and the column which look like the guide row and column are headed by the same element is a group property. In Section 5 you will be asked to prove this from the seven properties we have listed.

## Exercises

**1.** Explain why the following tables are *not* group tables. (If one of the seven properties is missing the table does not specify a group. In some cases it is possible to find more than one property which is missing.)

(a)

| | *A* | *B* |
|---|---|---|
| *A* | *A* | *A* |
| *B* | *A* | *A* |

(b)

| | *A* | *B* | *C* |
|---|---|---|---|
| *A* | *B* | *A* | *C* |
| *B* | *A* | *C* | *B* |
| *C* | *C* | *B* | *A* |

(c)

| | *A* | *B* | *C* |
|---|---|---|---|
| *A* | *A* | *B* | *C* |
| *B* | *B* | *C* | *D* |
| *C* | *C* | *D* | *E* |

(d)

| | *A* | *B* | *C* |
|---|---|---|---|
| *A* | *A* | *B* | *C* |
| *B* | *C* | *A* | *B* |
| *C* | *B* | *C* | *A* |

(e)

| | *A* | *B* | *C* |
|---|---|---|---|
| *A* | *A* | *C* | *B* |
| *B* | *B* | *A* | *C* |
| *C* | *C* | *B* | *A* |

(f)

| | *A* | *B* | *C* | *D* | *E* |
|---|---|---|---|---|---|
| *A* | *A* | *B* | *C* | *D* | *E* |
| *B* | *B* | *C* | *A* | *E* | *D* |
| *C* | *C* | *E* | *D* | *A* | *B* |
| *D* | *D* | *A* | *E* | *B* | *C* |
| *E* | *E* | *D* | *B* | *C* | *A* |

**2.** Consider three equally spaced positions *a, b, c* on a circle and three objects *X, Y, Z* at these three positions. If we allow only rearrangements which move all these objects in a clockwise direction, there are three possible rearrangements. What are they? Let product mean one rearrangement followed by a second. Find the table of products and show that it has all seven group properties. How is this table related to that of Group 2?

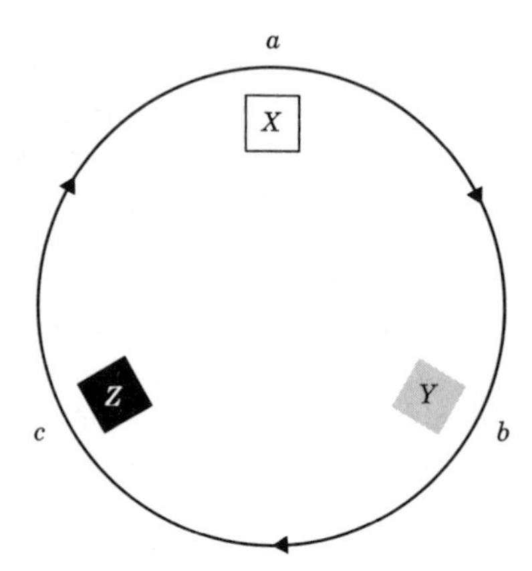


**3.** Repeat Exercise 2 using four equally spaced positions on a circle. (For Property 7 check only a few cases.) How is this group related to Group 3?

**4.** Consider a set of two objects $\{E, O\}$. Think of $E$ as standing for even and $O$ for odd. Make a table for $\{E, O\}$ if the operation indicates whether the sum of the various combinations of even and odd are even or odd; for example $E + O = O$. Does this table specify a group? Is the table like that of any group we have already considered?

**5.** Imagine a "clock" having the numbers 0, 1, 2, 3, 4, 5, 6 equally spaced on its face. This clock can be used to find the "sum" of any two of the numbers 0, 1, 2, 3, 4, 5, 6. The "sum" $2 + 3$ is obtained by placing a hand of the clock at 2 and moving *three* numbers in a clockwise direction, so $2 + 3 = 5$. Similarly, $4 + 3 = 0$, $2 + 6 = 1$, and $4 + 1 = 5$. Write out the table of all "sums" formed from the set 0, 1, 2, 3, 4, 5, 6. Is this a group table? What common real-life situation uses this "clock arithmetic"?

***6.** The following definition is the one most commonly used for a (not necessarily finite) group.

A *group* is:

I. A *set* of objects **G** together with an *operation* on these objects, that is, for any two objects $x$ and $y$ in **G** there is a unique product $xy$ which is also an object in **G**.

II. **G** contains an *identity* element $i$ such that for all $x$ in **G**, $xi = x = ix$.

III. For every element $x$ in **G** there exists an *inverse* element $x'$ in **G** such that $xx' = i = x'x$.

IV. The operation is *associative,* that is, for all $x$, $y$, $z$ in **G**, $(xy)z = x(yz)$.

For finite sets this definition specifies the same type of mathematical system as the definition of a group given in this section. Let **G** be a set with a table satisfying Properties 1 to 7 and prove the following:

(a) **G** has Property I. (Hint: Use Properties 1 to 3.)

(b) If the column which looks like the guide column is headed by $i$, then for all $x$ in **G**, $xi = x$.
If the row which looks like the guide row is headed by $j$, then for all $x$ in **G**, $jx = x$.
(It can also be shown using Properties 5 and 6 that $i = j$. This proof will be asked for in Exercise 4 of Section 5. Combining these facts proves that Property II is true for **G**.)

(c) **G** has Property III. (Assume Property II has been proved and the identity is $i$.)

(d) Property IV is true for **G**.

***7.** The following familiar mathematical systems satisfy the definition of a group given in Exercise 6. In each example name the identity and specify the inverse $x'$ of each element $x$.

(a) The integers with ordinary addition.

(b) The rationals (all fractions) with ordinary addition.

(c) All the rationals except 0 with ordinary multiplication. (Why is it necessary to omit 0?)

(d) The real numbers (all decimal numbers) with ordinary addition.

(e) All real numbers except 0 with ordinary multiplication.

***8.** Explain why the following are not infinite groups:

(a) The integers with ordinary multiplication.

(b) The natural numbers with ordinary addition.

(c) The set of all proper fractions (fractions in which the numerator is less than the denominator) with ordinary addition.

(d) The natural numbers with multiplication.

## 3. SUBGROUPS

There is another thing that we might have noticed about Groups 1 to 3. In each table the upper left hand quarter of the table is a smaller group table. This means that a subset (part) of the set with the products of the original table has all of the seven group properties, that is, it is a group. If a subset of a group is also a group using the original products, then we call this subset a *subgroup* of the original group.

***Show that* {A, B, C} *with the products of Group 2 is a subgroup of Group 2.* ({A, B, C} *is read "the set of elements* A, B, C.")**

If we form the table of products for $\{A, B, C\}$ we get:

| | $A$ | $B$ | $C$ |
|---|---|---|---|
| $A$ | $A$ | $B$ | $C$ |
| $B$ | $B$ | $C$ | $A$ |
| $C$ | $C$ | $A$ | $B$ |

The first six group properties are apparent in the table. We will not have to check the seventh property, associativity, because the products are the same as in the original large table. Since we are assuming the large table is a group, these products must have the associative property. We might say that they inherit associativity from the original group.

***Show that* {A, F} *with the products of Group 2 is a subgroup of Group 2.***

The subset $\{A, F\}$ is a subgroup because that table has all seven group properties.

| | $A$ | $F$ |
|---|---|---|
| $A$ | $A$ | $F$ |
| $F$ | $F$ | $A$ |

***Show that* {A, B} *with the products of Group 2 is not a subgroup of Group 2.***

Since $BB = C$ in Group 2, the table for $\{A, B\}$ using the products of the original group would not have Property 3.

***Show that* {A} *with the product of Group 2 is a subgroup of Group 2.***

The subset $\{A\}$ is a subgroup because the table has all seven group properties.

| | $A$ |
|---|---|
| $A$ | $A$ |

***Show that* {A, B, C, D, E, F} *with the products of Group 2 is a subgroup of Group 2.***

Since in mathematics we consider an entire set to be a subset of itself, this is a subset of Group 2. The subset {*A, B, C, D, E, F*} with the products of Group 2 is simply Group 2 and so has all seven properties.

***Find all the subgroups of Group 1.***

Group 1 has two subgroups:

| | A |
|---|---|
| A | A |

and the whole group.

***Find all the subgroups of Group 2.***

To find all the subgroups of Group 2, we must test every subset of {*A, B, C, D, E, F*} to see whether or not it is a group when the products of Group 2 are used. Only six of these subsets have the seven properties:

| | A |
|---|---|
| A | A |

,

| | A | D |
|---|---|---|
| A | A | D |
| D | D | A |

,

| | A | E |
|---|---|---|
| A | A | E |
| E | E | A |

,

| | A | F |
|---|---|---|
| A | A | F |
| F | F | A |

,

| | A | B | C |
|---|---|---|---|
| A | A | B | C |
| B | B | C | A |
| C | C | A | B |

,

and the whole group.

***Find some of the subgroups of Group 3.***

Group 3 has thirty different subgroups. Some of them are:

| | A |
|---|---|
| A | A |

;

| | A | J |
|---|---|---|
| A | A | J |
| J | J | A |

,

| | A | K |
|---|---|---|
| A | A | K |
| K | K | A |

,

and seven more subgroups of two elements;

| | A | B | C |
|---|---|---|---|
| A | A | B | C |
| B | B | C | A |
| C | C | A | B |

and three more subgroups of three elements;

| | A | M | R | K |
|---|---|---|---|---|
| A | A | M | R | K |
| M | M | A | K | R |
| R | R | K | A | M |
| K | K | R | M | A |

and six more subgroups of four elements.

Group 3 also has four subgroups of six elements, three subgroups of eight elements, one subgroup of twelve elements, and the entire group is a subgroup of itself.

We have come a long way from considering a few books on a shelf. The physical rearrangement of books led us to the abstract concept of a group, a mathematical system in which we can multiply without numbers. Groups also contain subgroups. Considering just four books led us to a group of twenty-four elements, which contains thirty subgroups—an amazing number of mathematical systems to be associated with just four books.

## Exercises

1. Find the remaining seven two-element subgroups of Group 3.
2. Find the remaining three three-element subgroups of Group 3.
3. In Group 3, show that $\{A, S, J, T, N, Q, K, L\}$ is a subgroup.
4. The table for the subgroup in the last exercise can be thought of simply as a group table, so it, too, may have subgroups. Find all the subgroups of this table. (There are ten of them.)
5. Multiplication of numbers is commutative, that is, $AB = BA$. Does a group necessarily have this property? Why, or why not?

### 4. COSETS

Let's look at the group of eight elements discussed in Exercises 3 and 4 of Section 3. We'll call it Group 4. Its table is as follows:

| | A | S | J | T | N | Q | K | L |
|---|---|---|---|---|---|---|---|---|
| A | A | S | J | T | N | Q | K | L |
| S | S | J | T | A | K | L | Q | N |
| J | J | T | A | S | Q | N | L | K |
| T | T | A | S | J | L | K | N | Q |
| N | N | L | Q | K | A | J | T | S |
| Q | Q | K | N | L | J | A | S | T |
| K | K | N | L | Q | S | T | A | J |
| L | L | Q | K | N | T | S | J | A |

GROUP 4

The subgroups of Group 4 are: $\{A\}$, $\{A, J\}$, $\{A, N\}$, $\{A, Q\}$, $\{A, K\}$, $\{A, L\}$, $\{A, S, J, T\}$, $\{A, J, N, Q\}$, $\{A, J, K, L\}$, and the entire group itself.

From this example and the other examples of subgroups in the last section we should be able to make a conjecture about the relation between the number of elements in a group and the number of elements in any subgroup of the group.

***Can you formulate a conjecture?***

> CONJECTURE: The number of elements in a subgroup of a group is a divisor of the number of elements in the group.

This conjecture is true and is known as the Theorem of Lagrange. Let's see if we can find a proof of this theorem.

At present the only tools which we have to work with are the seven properties of a group table and the notion of a subgroup. To find a proof of the theorem we will have to step outside the statement of the problem and find a new tool to help us. This happens often in mathematics—we have to use information from outside the problem to solve it. It would be nicer if we didn't have to do this, but many times it is not possible to find a solution using only the terms of the problem itself. Although this procedure seems unfortunate at first, it may pay off in the long run. If we are lucky, the new tool which we introduce may prove valuable for attacking other problems.

The new tool which we need for the Lagrange Theorem is the idea of a *coset*. For any subgroup of a group, as $\{A, N\}$ above, we can consider several related sets which we can obtain by "multiplying" $\{A, N\}$ from the left by one element of the large group. Using the products indicated in the table for Group 4, the sets related to $\{A, N\}$ are:

$$
\begin{aligned}
A\{A, N\} &= \{AA, AN\} = \{A, N\},\\
S\{A, N\} &= \{SA, SN\} = \{S, K\},\\
J\{A, N\} &= \{JA, JN\} = \{J, Q\},\\
T\{A, N\} &= \{TA, TN\} = \{T, L\},\\
N\{A, N\} &= \{NA, NN\} = \{N, A\},\\
Q\{A, N\} &= \{QA, QN\} = \{Q, J\},\\
K\{A, N\} &= \{KA, KN\} = \{K, S\}, \text{ and}\\
L\{A, N\} &= \{LA, LN\} = \{L, T\}.
\end{aligned}
$$

These sets are called the *left cosets of* $\{A, N\}$ *relative to* Group 4 (because we are multiplying from the left by each of the elements of Group 4).

***Find the left cosets of* {A, J}.**

The left cosets of $\{A, J\}$ are: $A\{A, J\} = \{A, J\}$, $S\{A, J\} = \{S, T\}$, $J\{A, J\} = \{J, A\}$, $T\{A, J\} = \{T, S\}$, $N\{A, J\} = \{N, Q\}$, $Q\{A, J\} = \{Q, N\}$, $K\{A, J\} = \{K, L\}$, and $L\{A, J\} = \{L, K\}$.

***Find the left cosets of* {A, J, K, L}.**

The left cosets of $\{A, J, K, L\}$ are: $A\{A, J, K, L\} = \{A, J, K, L\}$, $S\{A, J, K, L\} = \{S, T, Q, N\}$, $J\{A, J, K, L\} = \{J, A, L, K\}$, $T\{A, J, K, L\} = \{T, S, N, Q\}$, $N\{A, J, K, L\} = \{N, Q, T, S\}$, $Q\{A, J, K, L\} = \{Q, N, S, T\}$, $K\{A, J, K, L\} = \{K, L, A, J\}$, and $L\{A, J, K, L\} = \{L, K, J, A\}$.

***The left cosets of* {A, N}, {A, J}, *and* {A, J, K, L} *have some common properties. Can you find any?***

In each of these examples the following are true:

A. Every left coset has the same number of elements in it as the subgroup being considered.
B. Every element of the large group belongs to some left coset.
C. Two left cosets are either identical or have no element in common. (Since we are concerned only with what is in a coset and not concerned about the order in which the elements are listed, then $\{A, J, K, L\} = \{J, A, L, K\}$, and similarly for the other cosets.)

To prove the Theorem of Lagrange all we will have to do is to show that these three things always happen when we consider the left cosets of a subgroup. If they do, we can get the whole group by taking the *union* of the distinct cosets because every element in the group is in some left coset. (The union of two sets $X$ and $Y$ is the set of all elements in $X$ or in $Y$ or in both.) Group 4 is the union of $\{A, N\}$, $\{S, K\}$, $\{J, Q\}$, and $\{T, L\}$ in the first example, $\{A, J\}$, $\{S, T\}$, $\{N, Q\}$, and $\{K, L\}$ in the second example, and $\{A, J, K, L\}$ and $\{S, T, Q, N\}$ in the third. Since

1. every coset has the same number of elements as the subgroup,
2. the union of some of the cosets is the whole group, and
3. the distinct cosets have no elements in common,

it follows that the number of cosets multiplied by the number of elements in each coset is the number of elements in the group. Therefore, the number of elements in the subgroup must divide the number of elements in the group. (In fact, the number of elements in the group divided by the number of elements in the subgroup is the number of distinct left cosets.)

In the next section we will prove that Properties A to C are true for any set of left cosets of a subgroup. Then we will explain again why the Theorem of Lagrange follows from these properties.

## Exercises

1. Find the left cosets of $\{A, L\}$ relative to Group 4.
2. Find the left cosets of $\{A, S, J, T\}$ relative to Group 4.
3. Show that Properties A to C are true for the cosets of Exercise 1. Do the same for the cosets of Exercise 2.
4. How can Group 4 be obtained as the union of the left cosets of $\{A, L\}$?
5. How can Group 4 be obtained as the union of the left cosets of $\{A, S, J, T\}$?
6. How many distinct left cosets appear in Exercise 1? How is this number related to the number of elements in the subgroup $\{A, L\}$ and the number of elements in Group 4?
7. How many distinct left cosets appear in Exercise 2? How is this number related to the number of elements in the subgroup $\{A, S, J, T\}$ and the number of elements in Group 4?

### 5. THE THEOREM OF LAGRANGE

We will now show that Properties A to C of the last section always occur if we form the set of all left cosets of a group. These properties will help us prove the Theorem of Lagrange, so they are lemmas.

Because we will be speaking about *any* finite group, once again (as in Chapter 1) we will have to use general mathematical symbols. Let's let the set of the group be $\mathbf{G} = \{A_1, A_2, \ldots, A_n\}$, where $n$ is the number of elements in **G**. We will also have to talk about any subgroup of **G**, we'll call it **H**. We can symbolize **H** by $\{B_1, B_2, \ldots, B_d\}$ where every $B$ is an $A$ in **G** and $d \leq n$. The left cosets of **H** are $A_1 \{B_1, B_2, \ldots, B_d\}$, $A_2 \{B_1, B_2, \ldots, B_d\}, \ldots, A_n \{B_1, B_2, \ldots, B_d\}$. We could also write the left cosets as $A_1 \mathbf{H}, \ldots, A_n\mathbf{H}$. Sometimes to avoid using subscripts we will speak of an element $X$ in **G**. It is understood then that $X$ is one of the $A$'s. Similarly, we will speak of an element $D$ in **H**, and of course $D$ is one of the $B$s.

LEMMA A: If **H** is a subgroup of a group **G**, every left coset of **H** relative to **G** has the same number of elements as **H**.

*Proof:* Let $\mathbf{H} = \{B_1, B_2, \ldots, B_d\}$; **H** has $d$ distinct elements. Let $A_i\mathbf{H}$ be any left coset of **H** relative to **G**; $A_i\mathbf{H} = \{A_iB_1, A_iB_2, \ldots, A_iB_d\}$. This coset is a part of one of the rows of the table for **G**. Since every element of **G** appears only once in each row, the $d$ elements in $A_i\mathbf{H}$ are all distinct. This means that $A_i\mathbf{H}$ has as many elements as **H**. ■

LEMMA B: If **H** is a subgroup of **G**, then every element of **G** belongs to some left coset of **H** relative to **G**.
*Proof:* The table for group **G** has a column which looks like the guide column. Let's call the element at the head of this column $A_j$. Because the $A_j$ column looks like the guide column, for every element $X$ in **G**, $XA_j = X$. **H** is a subgroup of **G**, so **H** has a column which looks like its guide column. No element appears twice in a row of the **G** table, so the column of **H** which looks like its guide column is a part of the $A_j$ column in the **G** table. This means that $A_j$ is an element of **H**.

We are trying to show that any element $Y$ of **G** is in some left coset of **H**. We can find $Y$ in the coset $Y\mathbf{H} = \{YB_1, YB_2, \ldots, YB_d\}$ because one of the $B$s is $A_j$ and $YA_j = Y$. ■

LEMMA C: If **H** is a subgroup of **G** and $X\mathbf{H}$ and $Y\mathbf{H}$ are two left cosets, then either $X\mathbf{H}$ and $Y\mathbf{H}$ have no element in common or $X\mathbf{H}$ and $Y\mathbf{H}$ are identical.

*Proof:* Since $X\mathbf{H}$ and $Y\mathbf{H}$ are two sets, logic tells us that either they have no element in common, or else they have an element (or elements) in common. If they have no element in common the lemma is true. If they have an element in common, let's call this common element $C$. Since $C$ is in $X\mathbf{H}$, $C = XD$ for some $D$ *in* **H**; since $C$ is in $Y\mathbf{H}$, $C = YE$ for some $E$ in **H**; $C = XD$ and $C = YE$ implies that $XD = YE$.

As in Lemma B we know that $A_j$ is in **H** and $ZA_j = Z$ for all elements $Z$ in **G**. Every element of **H** appears in every row of the **H** table, so in particular $A_j$ is in the $D$ row of **H**. In order for $A_j$ to be in the $D$ row there must be another element $D'$ in **H** such that $DD' = A_j$. Using the equality we found in the last paragraph $XD = YE$, so $(XD)D' = (YE)D'$. Now applying the associative property of **G**, $Y(ED') = (YE)D' = (XD)D' = X(DD') = XA_j = X$. The fact, $X =$

$Y(ED')$, will help us show that the coset $X\mathbf{H}$ and $Y\mathbf{H}$ are identical.

Let $F$ be any element in $X\mathbf{H}$; we will show that $F$ is also in $Y\mathbf{H}$. If $F$ is in $X\mathbf{H}$ this means that $F = XK$ for some $K$ in $\mathbf{H}$. But $X = Y(ED')$ from above, so $F = [Y(ED')]\ K = Y[(ED')K]$. Since $E, D'$, and $K$ are all in $\mathbf{H}$, the product $(ED')K$ is an element of $\mathbf{H}$ and $F$ is one of the elements in the coset $Y\mathbf{H}$. This shows that every element of $X\mathbf{H}$ is an element of $Y\mathbf{H}$. Because $X\mathbf{H}$ and $Y\mathbf{H}$ have the same number of distinct elements, this means that they are identical. ■

THEOREM OF LAGRANGE: If **H** is a subgroup of a group **G**, then the number of elements of **H** is a divisor of the number of elements of **G**.

*Proof:* Using Lemma B we know that **G** is a union of some left cosets of **H**. Let's form **G** by taking a union of distinct cosets (no coset is used twice). Each of these cosets has as many elements as **H** by Lemma A—$d$ elements. The distinct cosets have no element in common by Lemma C. If we think of **G** as formed this way, we can see that **G** can be partitioned into distinct subsets of $d$ elements each, therefore, $d$ must divide the number of elements of **G**. ■

At this stage of our excursion into mathematics we probably find proofs like the ones we have just completed still a little difficult to understand. Even though all the experimental work building up to a proof is comprehensible, there is a sudden shift in the difficulty—mostly because of the symbolic language. The most important thing now is that we do not get discouraged. We may be learning the language even while we feel that no progress is being made. It would be dishonest for us to picture mathematics as only intuitive and experimental. A rigorous proof is the finished product the mathematician is after. For the time being, we may have to observe from the sidelines with something less than total comprehension. At this point we must simply try to understand as much as possible, and then remind ourselves that we are looking at a real mathematical proof and this is the level at which a proof must be carried out by a mathematician.

***You might enjoy the following experiment in mathematical learning. If the proofs were difficult for you, or even totally incomprehensible, after you have completed working through two more chapters, return to this one and read it again. You may be surprised to find that the material is much easier. How could you account for this?***

## Exercises

1. Let **G** be Group 4 (Section 4) and $\mathbf{H} = \{A, K\}$. Rewrite the proof of Lemma A in terms of this specific group and subgroup. (This exercise will not be a proof of Lemma A because it is a specific example; however, an example sometimes helps us understand the general proof.)
2. Does the group described in Exercise 5, Section 2, have any subgroups other than $\{0\}$ and $\{0, 1, 2, 3, 4, 5, 6\}$? Why, or why not?
3. Could a group containing exactly ten elements have any subgroups? If so, how many elements could be in a subgroup? Does the Lagrange Theorem guarantee that a group of ten elements has a subgroup of five elements?
4. Prove that in a group table, if the row that looks like the guide row is headed by guide element $X$, then the column which looks like the guide column is headed by $X$.

*5. Prove that in a group, if $XS = S$ for all elements $S$ of the group, then $SX = S$ for all elements $S$ in the group.

### 6. NORMAL SUBGROUPS

In the proof of the Lagrange Theorem we had to introduce the notion of the set of left cosets of a subgroup relative to a group. It seemed unfortunate that we had to step outside the statement of the theorem in order to find a proof. Actually this is not the case; the notion of coset is useful for other things besides the proof of the Lagrange Theorem. Let's look at another use now.

We have been forming the left cosets of a subgroup. We might wonder what would happen if we formed the *right* cosets of a subgroup; for example, the right cosets of $\{A, N\}$ are:

$$\begin{aligned}
\{A, N\}A &= \{AA, NA\} = \{A, N\},\\
\{A, N\}S &= \{AS, NS\} = \{S, L\},\\
\{A, N\}J &= \{AJ, NJ\} = \{J, Q\},\\
\{A, N\}T &= \{AT, NT\} = \{T, K\},\\
\{A, N\}N &= \{AN, NN\} = \{N, A\},\\
\{A, N\}Q &= \{AQ, NQ\} = \{Q, J\},\\
\{A, N\}K &= \{AK, NK\} = \{K, T\},\\
\{A, N\}L &= \{AL, NL\} = \{L, S\}.
\end{aligned}$$

If we compare the right and left cosets of $\{A, N\}$ we see that the corresponding cosets are not always the same, $S\{A, N\} = \{S, K\} \neq \{S, L\} = \{A, N\}S$ and $T\{A, N\} = \{T, L\} \neq \{T, K\} = \{A, N\}T$.

***Find the right cosets of* {A, J}.**

In the case of $\{A, J\}$ each right coset is equal to the corresponding left coset:

$$
\begin{aligned}
A\{A, J\} &= \{A, J\} = \{A, J\}A, \\
S\{A, J\} &= \{S, T\} = \{A, J\}S, \\
J\{A, J\} &= \{J, A\} = \{A, J\}J, \\
T\{A, J\} &= \{T, S\} = \{A, J\}T, \\
N\{A, J\} &= \{N, Q\} = \{A, J\}N, \\
Q\{A, J\} &= \{Q, N\} = \{A, J\}Q, \\
K\{A, J\} &= \{K, L\} = \{A, J\}K, \text{ and} \\
L\{A, J\} &= \{L, K\} = \{A, J\}L.
\end{aligned}
$$

This indicates that all subgroups are *not* of the same nature; some subgroups have this extra property that their left cosets are equal to the corresponding right cosets—these subgroups are called *normal subgroups.*

***Show that* {A, S, J, T} *is a normal subgroup of Group 4.***

We should form all the left cosets of $\{A, S, J, T\}$ and all the right cosets and show that the corresponding cosets are equal.

***Show that* {A, K} *is not a normal subgroup of Group 4.***

The left coset $N\{A, K\} = \{N, T\}$ and the corresponding right coset $\{A, K\}N = \{N, S\}$, so $\{A, K\}$ is *not* a normal subgroup.

The normal subgroups of Group 4 are $\{A\}$, $\{A, J\}$, $\{A, S, J, T\}$, $\{A, J, N, Q\}$, $\{A, J, K, L\}$, and the whole group. Since we are always looking for relationships, we might ask if the normal subgroups of a group are related to each other in some way.

***Find a relationship between the normal subgroups of Group 4.***

If we consider the normal subgroups of a group, we see that some normal subgroups are subgroups of other normal subgroups, and some are not. We can use the relation "is a subgroup of" to *order* the normal subgroups in a lattice. The normal subgroups of Group 4 form the following lattice:

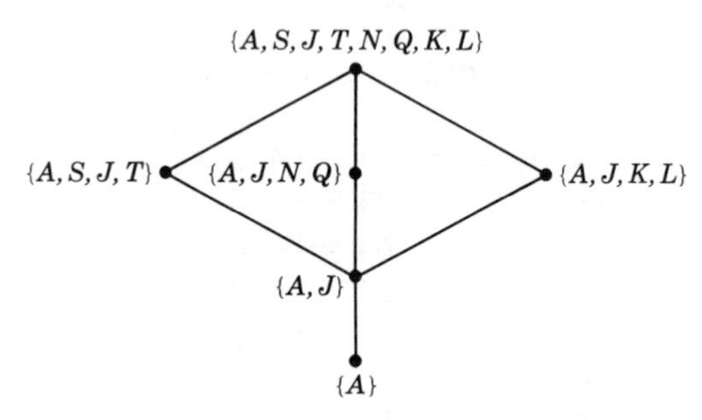


We should be a little surprised at this point that we have been able to find a lattice of normal subgroups. After all, our first experience with lattices occurred when we were studying divisors of numbers. Divisors and normal subgroups are two different concepts in mathematics and we would not expect them to have so much in common. It is, however, a very pleasant surprise. It seems to indicate that there is a basic unity about mathematics—similar structures appear even in unrelated parts of mathematics.

## Exercises

1. Show that $\{A\}$, $\{A, J\}$, $\{A, S, J, T\}$, $\{A, J, N, Q\}$, $\{A, J, K, L\}$, and Group 4 are all normal subgroups of Group 4. Show that Group 4 has no other normal subgroups.
2. Use Exercise 1 to make a conjecture about a subgroup of a group which has exactly half as many elements in it as the group. Prove or disprove your conjecture.
3. Find all the normal subgroups of Group 2 in Section 2. Find the lattice of these normal subgroups.
4. The set $\{A, B, C, D, E, F\}$ with this table is a group. It has four subgroups. Find them. Show that all the subgroups are normal. Draw the lattice of these normal subgroups.

| | A | B | C | D | E | F |
|---|---|---|---|---|---|---|
| A | A | B | C | D | E | F |
| B | B | C | D | E | F | A |
| C | C | D | E | F | A | B |
| D | D | E | F | A | B | C |
| E | E | F | A | B | C | D |
| F | F | A | B | C | D | E |

**5.** In the last exercise, how could we have predicted that all the subgroups would be normal without calculating the left and right cosets?

**6.** In the lattice of normal subgroups of Group 4, which subgroup is the largest subgroup of both $\{A, S, J, T\}$ and $\{A, J, N, Q\}$? Notice that it consists of all the elements which these two subgroups have in common. Which subgroup consists of the intersection (common elements) of $\{A, J, N, Q\}$ and $\{A, J\}$? Of $\{A, S, J, T, N, Q, K, L\}$ and $\{A, J, K, L\}$? How would you describe the position in the lattice of the intersection (largest common subgroup) of two normal subgroups?

**7.** If **H** and **K** are two subgroups, let **HK** be the set of all possible products $XY$ in which $X$ is an element of **H** and $Y$ is an element of **K**. In Group 4, what is **HK** if $\mathbf{H} = \{A, J\}$ and $\mathbf{K} = \{A, J, K, L\}$? If $\mathbf{H} = \{A\}$ and $\mathbf{K} = \{A, J, N, Q\}$? If $\mathbf{H} = \{A, S, J, T\}$ and $\mathbf{K} = \{A, J, N, Q\}$? How would you describe the position in the lattice of this "product" of two normal subgroups?

**8.** In Group 4, let $\mathbf{H} = \{A, K\}$ and $\mathbf{K} = \{A, N\}$. The subgroups **H** and **K** are not normal (Exercise 1). Form **HK** as in the last exercise. Is **HK** a subgroup of Group 4? Can you give a reason for using only the normal subgroups of a group to form a lattice?

## 7. WHAT HAVE WE LEARNED FROM GROUPS?

Our excursion into groups brings out several facts about mathematics. First of all, it shows us that *mathematics need not be about numbers.* Mathematics can deal with objects which act similarly to numbers. In our example we saw that rearrangements of books have some of the properties of numbers.

The proof of the Lagrange Theorem teaches us something about the way a mathematician works. As was pointed out as we developed the proof, he often has to introduce a *new tool* (cosets) from outside of the statement of the problem in order to solve the problem (the Lagrange Theorem). He then applies this new tool (cosets) to other problems (distinguishing normal subgroups).

Normal subgroups also illustrate a common phenomenon in mathematics. Normal subgroups of a group formed a lattice, just as all of the divisors of a number (in Chapter 1) formed a lattice. Very often the *same structure* appears in two unrelated parts of mathematics.

The systematic study of groups began toward the end of the 18th century. The work was given a big push by the results of Galois (1811–1832). Galois was studying the rearrangements of roots of equations of higher degrees (very much like our rearrangements of books). In the 20th century the theory of groups came into its own as a part of mathematics. Today the notion of group is one of the most basic concepts in mathematics.

In almost every branch of mathematics there are structures which have the properties of a group.

It has been found that there are phenomena in the physical world which also have a group structure. Groups are used to order and classify the atomic spectra. Groups are used in the special theory of relativity, which comes into play in atomic reactors and accelerators. Recently, group theory has been used to explain the variety of "fundamental" particles.

A mathematical idea often begins with some concrete situation (rearrangements of books, or of roots). The properties involved are then studied in an *abstract* way (for example, the properties of tables). Sometimes it is possible to *apply* this knowledge to other parts of mathematics, and sometimes it can even be applied to a *practical problem.* If a physical application is made, we have come full circle:

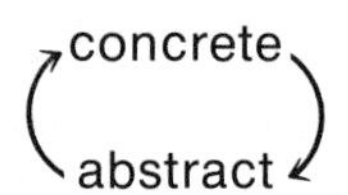


Even if there seems to be no application of the abstract mathematics to the real world, the mathematician still feels that his work is worthwhile. Perhaps an application will be found at a later date. If not, the abstract theory itself has a beauty about it. It required imagination and inspiration in its conception; it is precise, logical, interrelated. It is very much a *work of art,* and as such requires no further justification.

## Exercises

**1.** Imagine this sketch to be that of a water molecule. Interchanging the position of the hydrogen atoms leaves the molecule essentially the same. Call this

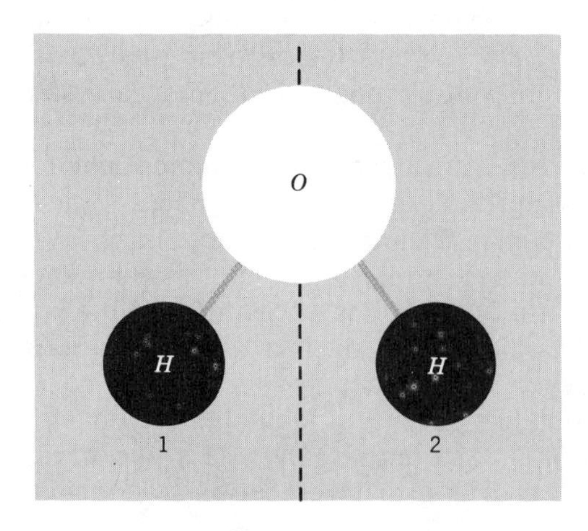

change of position $R$ since it can be thought of as a *reflection* in a vertical line through the oxygen atom. A change of position which leaves a figure looking the same as it did originally is called a *symmetry* of the figure. Represent no change by $I$, and call it a symmetry, too. Find the table of products of the two symmetries $I$ and $R$ in which product means one symmetry followed by another. Is this a group table?

***2.** Find the table of symmetries of the benzene molecule given here. (Assume all

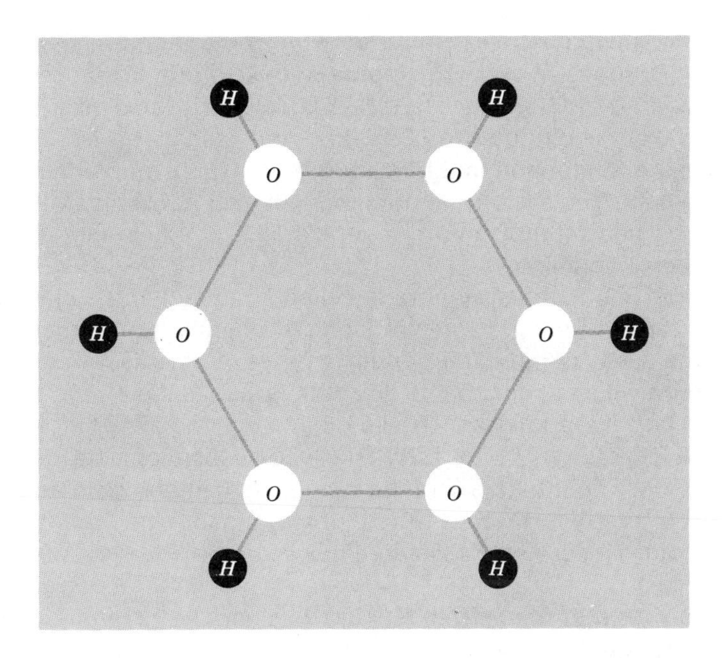


the bonds are the same.) There are twelve symmetries, six of them are reflections and six rotations. Is the table of symmetries that of a group?

**3.** Read one of the following:

Eddington, Sir Arthur Stanley, "The Theory of Groups," in *The World of Mathematics,* Vol. 3, J. R. Newman, Simon and Schuster, New York, 1956, pp. 1558–1573.

Keyser, Cassius J., "The Group Concept," in *The World of Mathematics,* Vol. 3, J. R. Newman, Simon and Schuster, New York, 1956, pp. 1537–1557.

Lieber, Lillian R., *Galois and the Theory of Groups,* The Science Press Printing Co., Lancaster, Pa., 1932.

Sarton, George, "Évariste Galois," *Osiris,* Vol. 3, 1937, pp. 24–259.

Sarton, George, "Lagrange's Personality," *Proceedings of the American Philosophical Society,* Vol. 88, 1944, pp. 457–496.

The article on Groups in a recent edition of the *Encyclopaedia Britannica.*

## 8. WHAT IS MODERN ALGEBRA?

Modern algebra, the branch of mathematics to which the theory of groups belongs, is also called higher algebra, or abstract algebra. It is different from the algebra we learned in high school in its abstractness. The elementary algebra of high school deals with numbers or symbols which represent numbers. In higher algebra we study abstract mathematical structures in which there are operations which resemble addition or multiplication.

Groups are mathematical structures with one operation; we called the operation a product. In modern algebra we also study structures with two operations, an "addition" and a "multiplication." Some of these structures are called rings, integral domains, and fields. The set of integers with ordinary addition and multiplication is an example of a ring and of an integral domain. The set of real numbers (all decimal numbers) with ordinary addition and multiplication is an example of a field (it is also a ring and an integral domain).

Another part of abstract algebra is called linear algebra. Linear algebra deals with systems which have two kinds of objects, an operation on one of them, and a way of combining the two types of objects. For example, in linear algebra we can consider as one set of objects the binomials $ax + by$ in which $a$ and $b$ are integers. These objects can be added; for example, $(2x + 3y) + [3x + (-1)y] = 5x + 2y$. The second set of objects might be the set of integers. The integers and the binomials can be combined by multiplication, $5(3x + 4y) = 15x + 20y$. In linear algebra we are interested in properties of systems which resemble this example (the systems are called vector spaces).

In 1945 a new part of abstract algebra was begun—categorical algebra. The structure studied is a collection of similar mathematical systems—a category. For example, the collection of all groups and the relationships (homomorphisms) between these groups is the category of groups. Other examples of categories are the category of rings and the category of fields. Although we will not attempt to describe categorical algebra in any detail, we can see even from this brief description that it is still another level of abstraction above that of a group. This is typical of mathematics, we seek a higher level of abstraction. The reason is economy. If an algebraist is able to prove something about categories in general, he gets information which will apply to every individual category.

Sometimes people think that all of mathematics was known to the Greeks and we have simply been reformulating their knowledge for over twenty centuries. This is far from the truth. Very little of our present day mathematics was known by the ancient Greeks. New mathematics is being discovered every day. Before 1945 the notion of a category did not exist. Today categorical algebra is a very lively area of research.

## Exercises

**READ SOME OF THE FOLLOWING**

Adler, Irving, *The New Mathematics,* A Mentor Book, The New American Library, New York, 1960.

Eilenberg, Samuel, "The Algebraization of Mathematics," *The Mathematical Sciences,* Edited by COSRIMS, M.I.T. Press, Cambridge, Mass., 1969, pp. 153–160.

McShane, E. J., "Vector Spaces and Their Applications," *The Mathematical Sciences,* Edited by COSRIMS, M.I.T. Press, Cambridge, Mass., 1969, pp. 84–96.

*Sawyer, W. W., "Algebra," *Scientific American,* Vol. 211, No. 3, September, 1964, pp. 70–78.

Sawyer, W. W., *A Concrete Approach to Abstract Algebra,* W. H. Freeman, San Francisco, 1959.

The article on Algebra (particularly the section on Modern Algebra) in a recent edition of the *Encyclopaedia Britannica.*

## Test

**1.** If $R = \begin{pmatrix} a & b & c \\ a & b & c \end{pmatrix}$, $S = \begin{pmatrix} a & b & c \\ b & c & a \end{pmatrix}$, and $T = \begin{pmatrix} a & b & c \\ c & a & b \end{pmatrix}$, find $ST$, $TT$, and $RS$.

**2.** Explain why each of the following is *not* a group table:

(a)

| | A | B |
|---|---|---|
| A | A | B |
| B | B | C |

(b)

| | A | B |
|---|---|---|
| A | B | B |
| B | B | B |

(c)

| | A | B | C | D |
|---|---|---|---|---|
| A | A | B | C | D |
| B | C | A | B | D |
| C | A | B | C | D |
| D | D | D | D | D |

**3.** This table is a group table.

| | R | S | T | U |
|---|---|---|---|---|
| R | R | S | T | U |
| S | S | R | U | T |
| T | T | U | R | S |
| U | U | T | S | R |

(a) Find all the subgroups of two elements.

(b) Is there any subgroup of three elements in this group? Why, or why not?

(c) Find all the left cosets of the subgroup $\{R, U\}$.

(d) Is the subgroup $\{R, U\}$ a normal subgroup? Why, or why not?

**4.** What outside reading have you done on modern algebra? Briefly summarize any material outside of this text which you found.

Mount Wilson & Palomar Observatory

3

# GEOMETRY

## Casting Shadows

**Man's attempts to understand the universe always involve some type of geometry.**

Most of us feel that mathematics, besides dealing with numbers, and systems resembling numbers, has something to do with geometrical figures. Our feeling is correct. Geometry is one of the oldest branches of mathematics. More correctly, the geometries form one of the oldest branches of mathematics; there are several geometries.

This multiplicity of geometries may come as a surprise for many of us because the only geometry we have seen so far is the Euclidean geometry of elementary and high school. We are in a position similar to that of the little boy who had a collie and never saw any other breed of dog. He began to think the word "dog" meant "collie," until one day he was told that a chihuahua was a dog. To understand geometry, we need to see a geometry different from Euclidean geometry. Let's look at *projective geometry.*

## 1. INTUITIVE PROJECTIVE GEOMETRY

Projective geometry was first used by the Renaissance artists. They were trying to portray three-dimensional scenes on two-dimensional canvases in a realistic way. They imagined that they were looking at the scene with one eye and that the rays of light from the scene were passing through a sheet of glass and converging at that eye. They placed their canvas where they had imagined the sheet of glass and then drew all the points where the rays would pass through the glass. This formed a two-dimensional picture which had many of the properties of the three-dimensional scene.

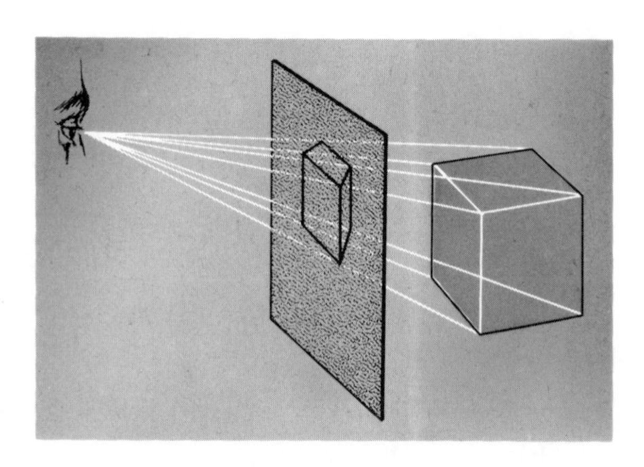

A similar thing occurs when shadows are cast. To keep the situation simple we will think about the shadows of two-dimensional objects. Imagine that a triangle is drawn with a grease pencil on a vertical sheet of glass. A light on one side of the glass will cast a shadow of the triangle on a horizontal surface. The shadow will usually be a triangle (provided the

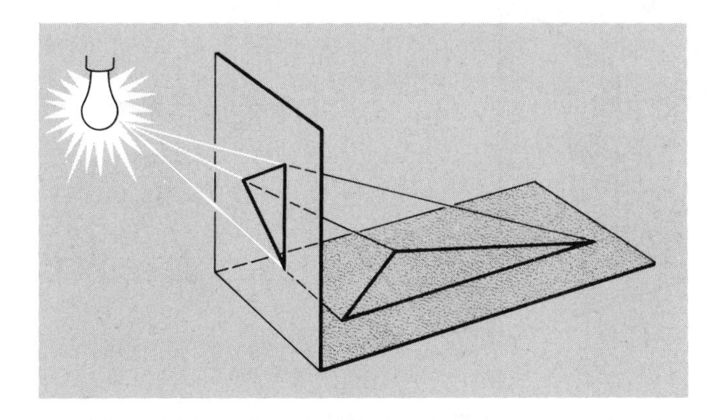

light is in the correct position to cast a shadow of the entire triangle). The original figure and its shadow may be different sizes and shapes, but they do have a common property—both are triangles. The property of being a triangle is preserved if we cast shadows.

We could draw various figures on the vertical glass, cast their shadows, and see which properties of the original figures are preserved by their shadows. As a temporary description, we'll say that *plane projective geometry is the study of the properties of two-dimensional figures preserved in their shadows.* In this chapter we will discuss only *plane* projective geometry.

***Imagine that you have drawn the following figures on a vertical sheet of glass. Describe the shadow of each figure and specify any common properties it has with the original figure:***

***a point (drawn as a dot),***
***a line segment,***
***a point on a line segment,***
***two intersecting line segments,***
***two perpendicular line segments,***
***two parallel segments,***
***an isosceles triangle,***
***a quadrilateral,***
***a parallelogram,***
***a square.***

It is hard to imagine casting the shadows of the figures listed above. If we try to help ourselves by drawing a picture on paper we also have some difficulty—the actual casting of shadows occurs in the real three-dimensional world and we must draw it on "two-dimensional" paper. It would be helpful if we could represent the casting of shadows in a simple way by a drawing on paper. There is a method which we will develop now.

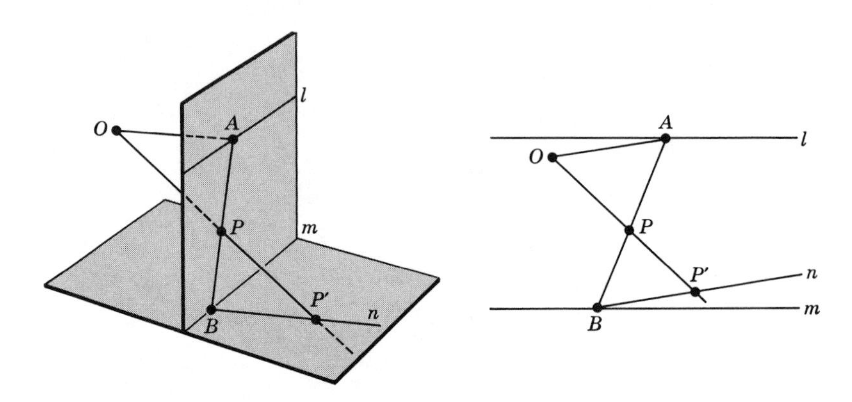


Imagine that we are standing (in the three-dimensional situation) behind the vertical plane and want to find the shadow cast on the horizontal plane of a point $P$ in the vertical plane. The light is at $O$ between us and the vertical plane. The line $l$ indicates the height of the light. The line $m$ is the line of intersection of the horizontal plane and the vertical plane. Let $P$ be a point in the vertical plane *between $l$ and $m$*. A two-dimensional representation of what we would see appears in the simplified drawing. The shadow $P'$ of $P$ in our simplified drawing is obtained as follows:

1. Draw any line through $P$ and not through $O$ cutting $l$ at $A$ and $m$ at $B$. (Use a straightedge.)
2. Draw $OA$ and $OP$ extended.
3. Draw the line $n$ through $B$ and parallel to $OA$. (A good approximation of the parallel is adequate. We are not doing mathematics at this point; we are experimenting with pictures.)
4. The intersection of $n$ and line $OP$ is the shadow $P'$.

***Find the shadow of the point* R.**

l

O •

R •

m

The shadow $R'$ of $R$ is found by the four steps listed above except that $R = P$. Again we must use a line $AB$ through $R$ which does not contain $O$.

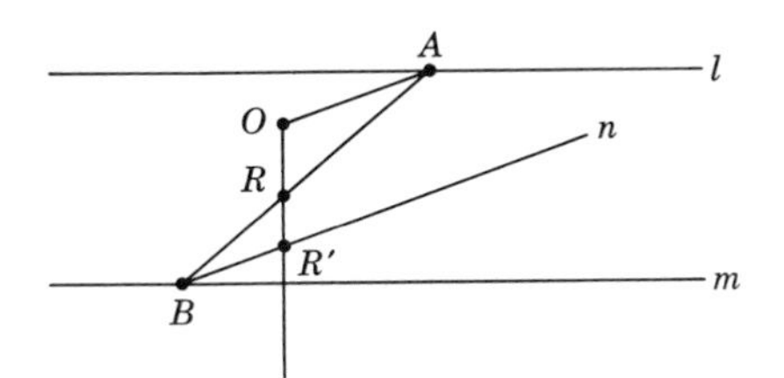


In both cases $P$ and $R$, we notice that the shadow of a point is a point.

In order to find the shadow of a segment, we first find the shadows of the endpoints of the segment. The line segment between the shadows of the endpoints is the shadow of the segment.

***Find the shadow of the segment* AB.**

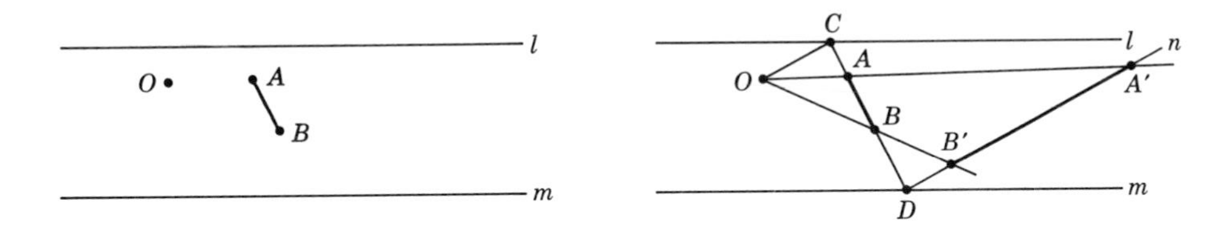


To find this shadow we can extend $AB$ until it cuts $l$ at $C$ and $m$ at $D$. We then draw $OC$, $OA$, and a line $n$ parallel to $OC$ and through $D$. The shadow of $A$ is $A'$, the point of intersection of $n$ and line $OA$. The shadow of $B$ is $B'$, the point of intersection of $n$ and line $OB$. The segment $A'B'$ is the shadow of the segment $AB$.

This construction seems to indicate that although the shadow of a segment is a segment, the length of a segment is not necessarily preserved by its shadow.

***Using our simplified method of drawing shadows, find the shadows of the figures in Fig. 3–1. It will sometimes be necessary to change the position of a figure to keep the "shadow" on your paper, however, we will always keep the figure between* l *and* m *and not touching* l *or* m.**

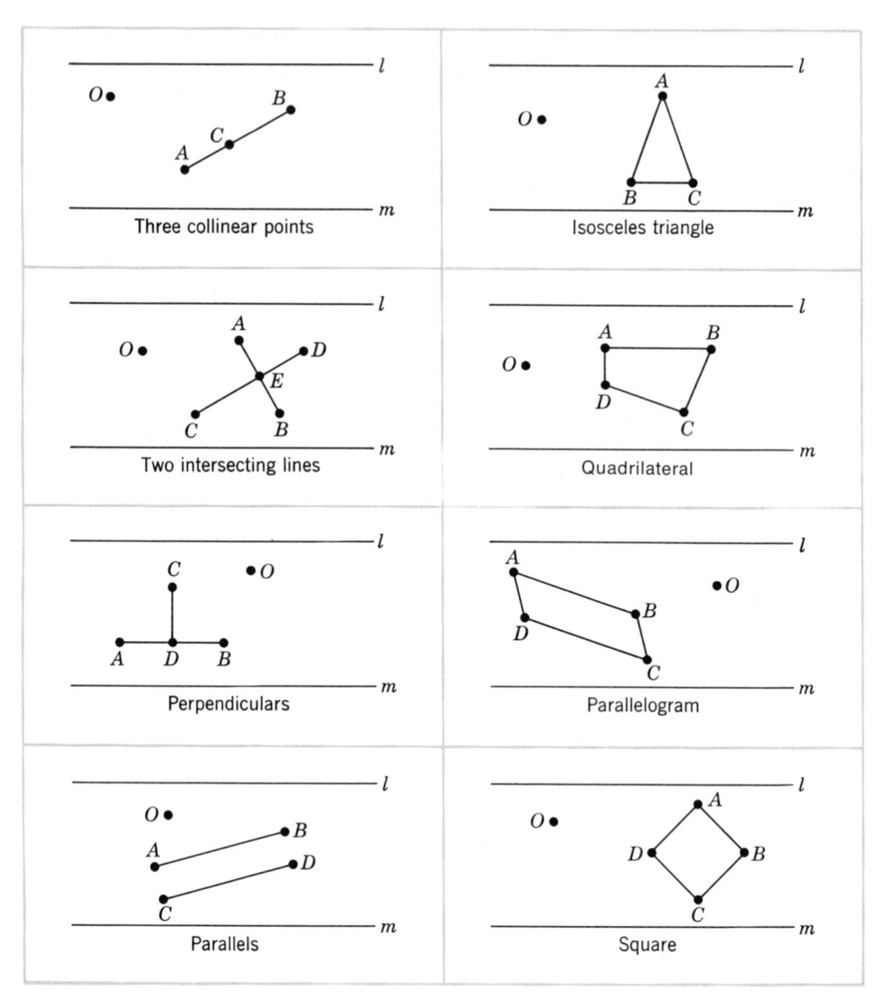


**Fig. 3–1.**

If we find the shadows of these figures correctly we can begin to see which properties might be studied in plane projective geometry. Points, line segments, triangles, and quadrilaterals all seem to have shadows which are the same type of object as the original figure. Of course, we haven't proved anything by drawing these pictures. We are merely getting a feeling for the kinds of properties that can be studied in a geometry which is the study of the properties of two-dimensional figures that are preserved in their shadows.

Strictly speaking, projective geometry deals with lines rather than line segments. However, we will speak of segments at the present time because we are only trying to build an intuitive idea of projective geometry. Segments will help us avoid some intuitive difficulties. Later we will give an accurate description of projective geometry.

If a point is on a segment, the shadow of the point is on the shadow of the segment. This was illustrated by the first drawing in the last experiment and means that "a point *being on* a segment" is a projective concept. The second drawing shows that the shadow of two intersecting segments is a pair of intersecting segments. This illustrates that "*intersecting segments*" is a projective concept.

The drawings also show that some properties are *not* projective. The size of an angle usually changes when we consider its shadow. Perpendicular segments need not have perpendicular shadows. Parallel segments need not have parallel shadows. An isosceles triangle may have a triangle which is not isosceles for its shadow. The shadow of a parallelogram and a square may be just quadrilaterals. These figures illustrate that size of angle, parallelism, and length of segment are not projective properties.

## Exercises

**1.** Find the shadow of three line segments which intersect at a common point. What is preserved?

**2.** Find the shadow of a trapezoid. What is preserved?

**3.** Find the shadow of a right triangle. What is preserved?

**4.** Find the shadow of a rhombus. What is preserved?

**5.** Find the shadow of a circle. Is "circle" a projective concept?

***6.** Prove that in the construction of the image of a point $P$, if a line $AB$ through $P$ and not through $O$ is used and leads to the shadow $P'$, then any other line $CD$ through $P$ and not through $O$ will lead to the same shadow $P'$. ($A$ and $C$ are on $l$; $B$ and $D$ are on $m$.)

***7.** Is it possible to find the shadow of a point $P$ if we allow $P$ to be above $l$? Below $m$? On $m$? On $l$? What difficulty would we have if we tried to find the shadow of an entire line?

## 2. THE THEOREM OF DESARGUES

We have described plane projective geometry as the study of the properties of figures preserved in their shadows. Casting shadows *destroys* many of the properties we study in Euclidean geometry: distance, size of angle, parallels, isosceles triangles, right triangles, trapezoids, parallelograms, rhombuses, and squares. It may seem at first that there is nothing left to study in projective geometry, but this is not the case. Projective geometry is a large field of study, as rich and intriguing as the more familiar Euclidean geometry. In order to understand how the few basic notions of projective geometry can lead to interesting results, let's look at a theorem of projective geometry.

Our experiments in finding shadows indicate that points, line segments, triangles, points lying on a common line (collinearity), and lines being through a common point (concurrency) can all be studied in projective geometry. Therefore in projective geometry we could consider the following pair of triangles:

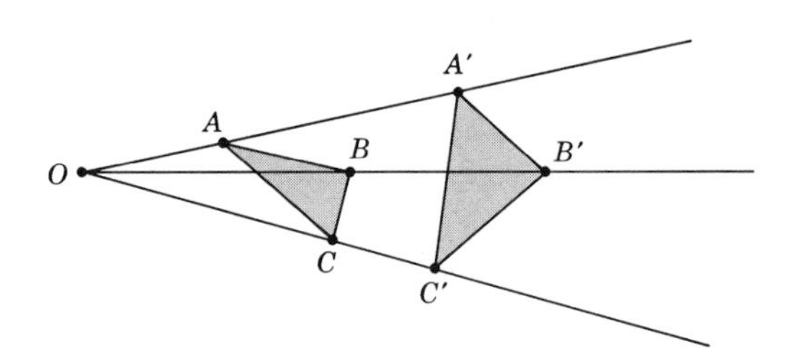


All of the points and lines in the figure are in one plane. It is also possible to think of them in space and build a three-dimensional projective geometry. We will keep the development simple by discussing only *plane* (two-dimensional) projective geometry. Triangles $ABC$ and $A'B'C'$ are such that the three lines joining the vertices $A$ and $A'$, $B$ and $B'$, $C$ and $C'$ all intersect at a common point $O$. We say that $AA'$, $BB'$, and $CC'$ are concurrent at $O$. Since these two triangles are related in a special way by concurrent lines we might wonder if there is any other relationship between the two triangles.

***Can you see any other relationship between triangles* ABC *and* A′B′C′*?***

The triangles are not the same size and not the same shape. In fact, these relationships would not be the kind we want to consider because they are not preserved by shadows. It is difficult to see any relationship besides the concurrency of $AA'$, $BB'$, and $CC'$. Perhaps if we would draw several more pairs of triangles $ABC$ and $A'B'C'$ with $AA'$, $BB'$, and $CC'$ concurrent at $O$, we might be able to see another relationship.

***Draw three more pairs of triangles satisfying the above condition.***

We might draw the following triangles:

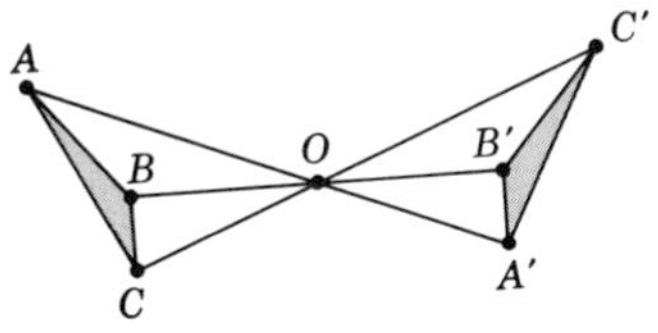

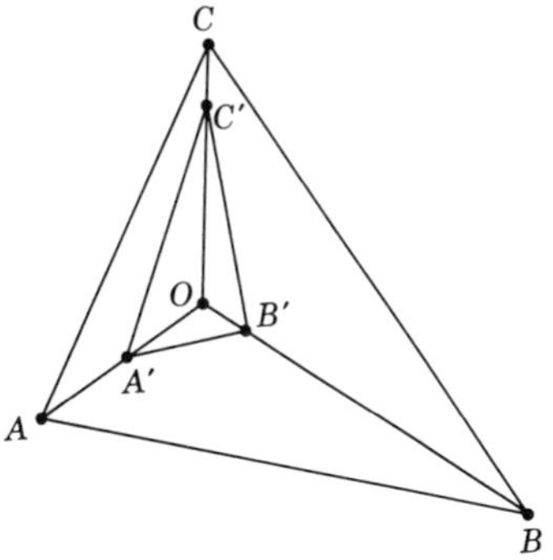

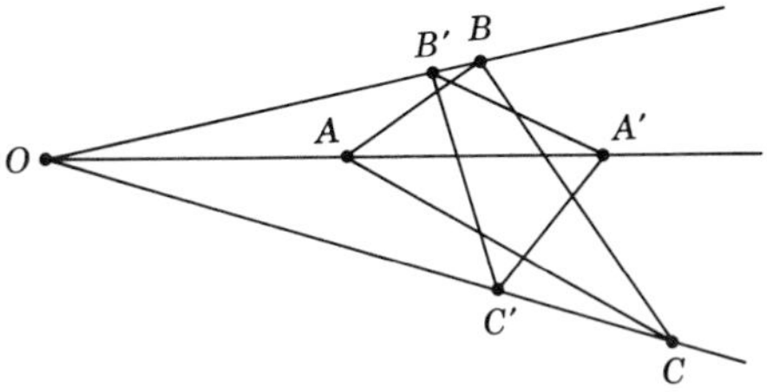

Each of these four pairs of triangles $ABC$ and $A'B'C'$ are related in the same way—$AA'$, $BB'$, and $CC'$ are concurrent at $O$. We are trying to see if they have any other property in common.

***In each of the four figures, extend the lines* AB, A′B′, BC, B′C′, AC *and* A′C′ *as in Fig. 3–2 on the next page. Can you see another relationship?***

For each of the pairs it looks like the intersections $D$, $E$, and $F$ of the corresponding sides extended are collinear ($D$ is the intersection of $AB$ and $A'B'$, $E$ of $BC$ and $B'C'$, and $F$ of $AC$ and $A'C'$.) From the four experiments we have just made, let's make a conjecture about certain triangles in projective geometry. The conjecture is true and is known as the *Two-dimensional Theorem of Desargues.*

> THEOREM OF DESARGUES: If two triangles (in a plane) have the lines joining their corresponding vertices concurrent and their corresponding sides or their extensions intersect, then these points of intersection are collinear.

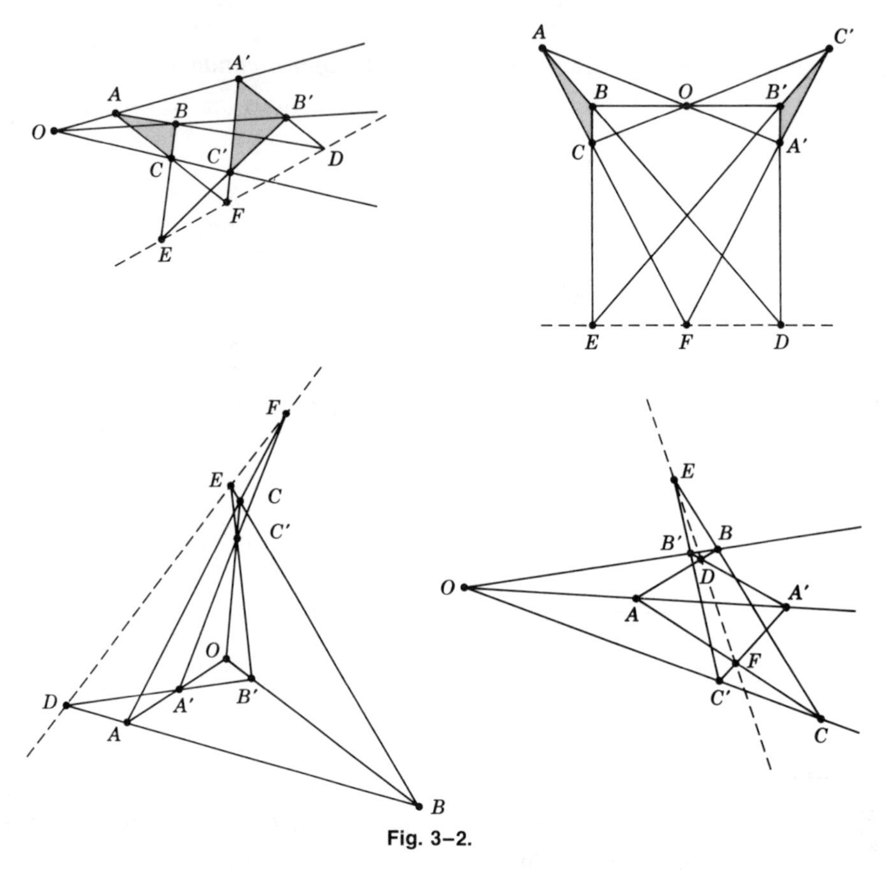


Fig. 3–2.

If we would draw more pictures the same thing would happen; however, that is not a proof of the Desargues Theorem. In order to be sure it will *always* happen, we will have to prove it from some accepted facts of projective geometry. A little later we will show one way that this can be done. The point of this section is merely to discover the theorem.

There are a few remarks that we should make about this statement of Desargues' Theorem. First of all, there is a three-dimensional form of the theorem which is also true. In the three-dimensional case the two triangles need not be in the same plane. Although the proof of the three-dimensional form is not very difficult, we will not consider it because in this chapter we are restricting ourselves to two-dimensional projective geometry.

Another remark that must be made concerns the requirement that the corresponding sides or their extensions intersect. As the theorem is worded, it is a theorem of Euclidean geometry. This is not surprising since the drawings we made on paper are Euclidean drawings. There is, of course, the possibility that the pairs of sides $AB$ and $A'B'$, $BC$ and $B'C'$, and $AC$ and $A'C'$ are each parallel. It is also possible that just one pair of corresponding sides is parallel. (It is not possible for only two pairs of corresponding sides to be parallel; see Exercise 4.) Parallelism as such is not studied in projective geometry. If projective geometry is approached through Euclidean geometry, there is a way to incorporate these parallel cases into the theorem and remove the restriction requiring that the corresponding sides intersect. We will briefly describe how this is done.

Historically, projective geometry was studied as a completion of the Euclidean plane. Lines instead of segments were considered, and a new line, called the *line at infinity,* was added to the Euclidean plane. A new set of points, called the *points at infinity,* were also added to the Euclidean plane. All of the points at infinity were on the line at infinity. Two parallel lines in the Euclidean plane are not parallel in the completion, that is, in the projective plane; instead, they intersect in one of the points at infinity. If three or more lines are all mutually parallel, they all intersect at the same point at infinity. In the projective plane the Theorem of Desargues reads as follows:

> THEOREM OF DESARGUES (PROJECTIVE FORM): If two triangles have the lines joining their corresponding vertices concurrent, then the points of intersection of the corresponding sides are collinear.

Considering the case mentioned above in which all three pairs of corresponding sides are parallel, the corresponding sides all intersect at points at infinity in the completed plane and these points are collinear since they are all on the line at infinity. Some of the other parallel cases will be discussed in the exercises.

Building projective geometry this way makes it clear that projective geometry is a generalization of Euclidean geometry. The Theorem of Desargues, which is true in Euclidean geometry, is true in a more general sense in projective geometry. It is possible to develop projective geometry independent of Euclidean geometry; we will do this in Section 4. In this section and the next one, we are approaching projective geometry as a generalization of Euclidean geometry in order to experiment and arrive at conjectures.

## Exercises

**1.** Draw three more examples of two triangles $ABC$ and $A'B'C'$ with $AA'$, $BB'$, and $CC'$ concurrent at $O$ (not a vertex) and in which the corresponding sides intersect at points $D$, $E$, and $F$. Are $D$, $E$, and $F$ collinear?

2. Draw two triangles $ABC$ and $A'B'C'$ with $AA'$, $BB'$, and $CC'$ concurrent at $O = A$. Does Desargues' Theorem apply to this case? Repeat the experiment two more times letting $O$ be a different vertex each time.
3. Why are our drawings not a proof of Desargues' Theorem? What is the value of the drawings?

*4. Draw two triangles $ABC$ and $A'B'C'$ such that $AB$ is parallel to $A'B'$, $BC$ is parallel to $B'C'$, and $AA'$, $BB'$, $CC'$ are concurrent. Make a conjecture about $AC$ and $A'C'$. Prove your conjecture using Euclidean geometry.

5. Draw two triangles $ABC$ and $A'B'C'$ such that $AB$ and $A'B'$ intersect at $D$, $BC$ and $B'C'$ at $E$, $AC$ and $A'C'$ at $F$, and $D$, $E$, $F$ are collinear. How are $AA'$, $BB'$, and $CC'$ related? Make a conjecture. How is your conjecture related to Desargues' Theorem?

*6. Consider two triangles $ABC$ and $A'B'C'$ in the Euclidean plane with $AA'$, $BB'$, and $CC'$ concurrent at $O$ (not a vertex) and in which $AB$ is parallel to $A'B'$, $BC$ and $B'C'$ intersect at a point $E$, and $AC$ and $A'C'$ intersect at a point $F$. Make a conjecture about $EF$. Assume that the Euclidean plane has been completed by the line at infinity as described at the end of this section. Why is the projective form of the Theorem of Desargues true for these triangles?

*7. Draw two triangles $ABC$ and $A'B'C'$ such that $AA'$, $BB'$, and $CC'$ are parallel and $AB$ and $A'B'$ intersect at $D$, $BC$ and $B'C'$ intersect at $E$, and $AC$ and $A'C'$ intersect at $F$. Do $D$, $E$, and $F$ seem to be collinear? Why is the projective form of Desargues' Theorem true for this case?

## 3. THE THEOREM OF PAPPUS

Let's see if we can find another projective theorem. The following figures consist of hexagons $ABCDEF$. A hexagon is composed of six sides (lines) and six vertices (points). The sides may or may not intersect in points other than the vertices. Only the intersections $A$, $B$, $C$, $D$, $E$, and $F$ are

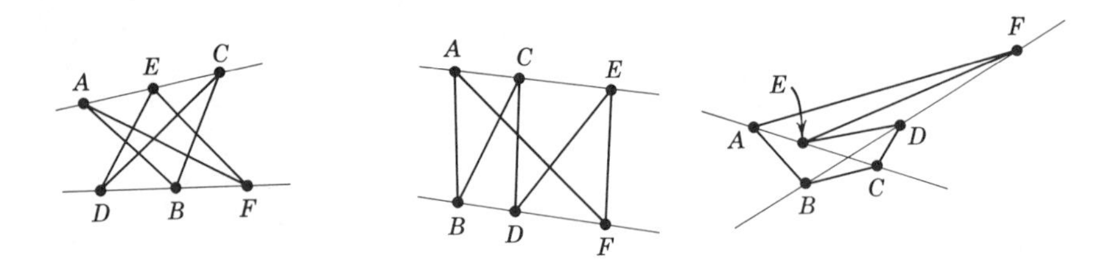


vertices. A hexagon can be studied in projective geometry because it is made up of lines and their intersections. All of the above hexagons have a common property.

***Can you see the common property of the hexagons?***

In each case, the hexagon has alternate vertices *A, C, E* and *B, D, F* lying on straight lines. Again we wonder if there is some other property that hexagons of this type have in common.

***Find another common property of the hexagons. (Hint: Find the intersections of the opposite sides* AB *and* ED, BC *and* EF, CD *and* AF.)**

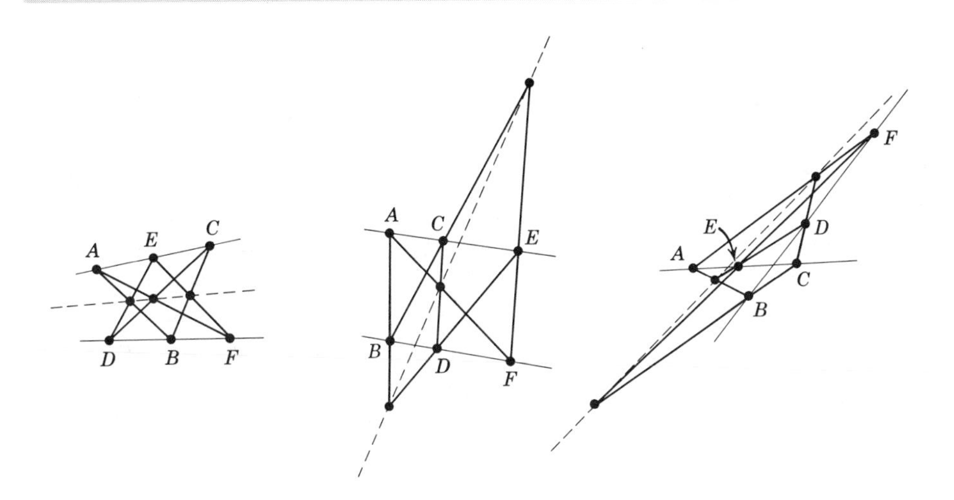


In all three cases, the intersections of the opposite sides are collinear.

***From the experiments above, make a conjecture about certain hexagons.***

The conjecture is true again this time; it is known as the Theorem of Pappus.

> THEOREM OF PAPPUS: If a hexagon has its alternate vertices on two straight lines and the opposite sides intersect, then the intersections of the opposite sides are collinear.

As in the Theorem of Desargues, we have drawn Euclidean pictures and the Theorem of Pappus as stated is a Euclidean theorem. It is possible that some of the pairs of opposite sides do not intersect, that is, they are parallel. If the line at infinity and all the points on it are added to the Euclidean plane to form a projective plane, then these parallel cases can be included in the theorem (see Exercise 2). The following is the more general projective form of the theorem.

> THEOREM OF PAPPUS (PROJECTIVE FORM): If a hexagon has its alternate vertices on two straight lines, then the intersections of the opposite sides are collinear.

## Exercises

**1.** Draw two more illustrations of the Theorem of Pappus.

***2.** Draw a hexagon *ABCDEF* such that *A, E, C* and *D, B, F* are collinear and such that *AF* is parallel to *CD,* and *FE* is parallel to *BC.* Make a conjecture about *AB* and *ED.* Why is the projective form of the Theorem of Pappus true for this hexagon?

**3.** Draw a hexagon *ABCDEF* such that the alternate sides *AB, CD* and *EF* meet at a point *O* and the other alternate sides *BC, DE,* and *FA* meet at another point *P.* What else seems to be true about this hexagon? (Hint: Look at *AD, BE,* and *CF.*) Repeat the experiment with a different hexagon satisfying the same conditions. Make a conjecture. How is this conjecture related to the Theorem of Pappus?

## 4. AXIOMS FOR PROJECTIVE GEOMETRY

We have described plane projective geometry as the study of the properties of figures which are unchanged by casting shadows. This approach could be refined and would lead to an accurate definition of projective geometry. All geometries can be described in a similar way as the study of *properties preserved under certain changes.*

This view of geometry is probably different from that most of us were exposed to in elementary or high school. There, geometry seemed to be the *content of certain statements.* For example, "All right angles are equal," and "Two points determine only one line," are facts about Euclidean geometry. All geometries can be thought of as the content of all of the true statements associated with the geometry.

Both of these approaches to geometry are legitimate. A geometer sometimes concentrates on the changes, at other times on the true statements.

Each approach has advantages and disadvantages. By studying geometry sometimes one way, sometimes another, a much fuller idea of the nature of geometry is developed.

At this point in our study of projective geometry it may be beneficial for us also to change our approach. Let's think of projective geometry now as the content of all of the true statements about projective geometry (that is, about the geometry which we already know about by studying shadows and by completing the Euclidean plane). The Theorem of Desargues and the Theorem of Pappus will be two of these statements. Besides these complicated statements, there are some simpler things that we know are true. "Two points lie on one and only one line," and "Two lines intersect in at most one point," would be among them.

We should notice at this point that a statement is *true* in projective geometry if it holds in the system we previously discussed. The word "true" does *not* imply anything about the real world.

If we want to consider *all* of the true statements about projective geometry, we will have a problem. There are too many true statements to simply list them. In order to deal with all of the statements we will have to organize them. The key to this organization is the fact that some of the true statements can be proved from others. If we can find a few statements from which all of the rest can be proved, then we will not have to list every true statement. Projective geometry would be completely characterized by the short basic list.

These basic statements are called *definitions* (if they introduce a new term) and *axioms* (if they are about terms previously introduced). Any statement which can be proved from the definitions and axioms is called a *theorem.* Notice that this use of "theorem" coincides with the way it was used in the first two chapters. (A theorem was a statement which could be proved from some basic facts.) The content of the *axiom system* (the collection of all the definitions, axioms, and theorems) is projective geometry.

Before we can state the definitions and axioms there is still another problem. The definitions and axioms are about some *objects* and *relations* between these objects. We must have some objects and relations to talk about before we can make any statements; these are called the *undefined terms.* Therefore, undefined terms are another essential part of an axiom system. We will assume that we have two sets of undefined objects—*points* and *lines.* We will also assume that there is an *undefined relation* of *being on* between certain points and lines. We will say things like "Point $P$ *is on* line $l$." Sometimes we will reverse the sentence and say "Line $l$ is on point $P$," or, equivalently, "Line $l$ goes through point $P$." Of course, the relation will not hold between all points and lines, so sometimes we will say "Point $Q$ is not on line $m$."

With these undefined notions of point, line, and being on (going through), we can define intersecting lines, collinear points, triangles, hexagons, and so on, in the usual way. The following axioms are often used for projective geometry.

A1. Two distinct points are on one and only one line.
A2. Two distinct lines go through a common point.
A3. There exist four points no three of which are on the same line.
A4. (Pappus) If a hexagon has its alternate vertices on two lines, then the points of intersection of the opposite sides are collinear.

We will assume the four axioms just listed. The third axiom is in our list to guarantee the existence of enough points to make the geometry interesting. A geometry of only two or three points would be so small that it would not have in any substantial way the properties which we want to be in projective geometry. (For example, we could not consider a hexagon.) Although A3 states that there are (at least) four points, we will soon see that there are really more than that in our projective geometry.

In A4, a hexagon is determined by any six points. If we name the points *A, B, C, D, E, F* and consider them in that order, that is, hexagon *ABCDEF*, then *A, C, E,* and *B, D, F* are the two sets of alternate vertices and the opposite sides are *AB* and *DE, BC* and *EF,* and *CD* and *AF.* The notion of alternate vertices and opposite sides does not depend on a picture which we draw, but rather on the order in which we consider the vertices.

The presence of A4, Pappus' Theorem, in our list of axioms may be a little confusing. Pappus' statement can be either an axiom or a theorem, depending upon our choice of the basic list of assumptions. We have chosen to use it as an axiom rather than as a theorem. We should notice also that we have omitted the requirement that the opposite sides intersect. It is no longer necessary to include this because the intersection is assured by A2.

We might wonder why the statement of A2 does not read "only one common point," and why Desargues' statement is not on the list. Both of these properties can be proved from the axioms, so in this development they are theorems rather than axioms.

Let's look at the first three axioms (we'll use the fourth one in the next section) and see if we can reason to some additional properties of projective geometry.

---

***Look at A1 and A3 and see if you can formulate a theorem about a line.***

---

We might make the following true conjecture:

THEOREM 1: There exists at least one line.

*Proof:* From A3 we know that four points exist. Using two of these points, by A1 there is a line through them. ■

---

***Look at A1 and A3 again and see if you can formulate a stronger theorem about the number of lines which exist.***

---

THEOREM 2: There exist at least six lines.

*Proof:* By A3 four points exist; let's call them *A, B, C,* and *D.* From A1, the pairs *A* and *B*, *A* and *C*, *A* and *D*, *B* and *C*, *B* and *D*, and *C* and *D* each lie on a line. No two of these lines can be the same, because then three of the points would be on the same line and this is not possible by A3. Therefore, there exist at least six lines. ■

We have already mentioned that we would expect the following statement to be a theorem.

THEOREM 3. Two distinct lines go through at most one point.

---

***See if you can prove Theorem 3. (Hint: Try a proof by contradiction.)***

---

*Proof of Theorem* 3: Assume that two lines $l$ and $m$, $l \neq m$, go through points $A$ and $B$, $A \neq B$. By A1 we know that $A$ and $B$ are on *only* one line; this contradicts $l$ and $m$ both going through $A$ and $B$. Therefore, our assumption was false and it is not possible that $l$ and $m$ go through two points. ■

---

***Try to formulate and prove some additional theorems.***

---

## Exercises

1. Prove that if $l$ is a line, then there exists a point not on $l$.
2. Prove that if $P$ is a point, then there is a line not through $P$.
3. Prove that any point has at least three lines going through it.
4. Prove that any line has at least three points on it.
5. Name three different hexagons determined by the six points $A, B, C, D, E, F$. In each case identify the alternate vertices and the opposite sides.

## 5. PAPPUS' THEOREM IMPLIES DESARGUES' THEOREM

In this section we will show that Desargues' Theorem can be proved from our list of axioms. Pappus' Theorem was already known at the time of Euclid. Desargues found his theorem during the first half of the 17th century. However, the relation between these two statements is a relatively new discovery. It was not discovered until 1905 that Desargues' Theorem could be proved from Pappus' Theorem.

We will give the steps in the proof of the Theorem of Desargues. It is more complicated than the proofs of the last section and must be worked through slowly.

***Give the reasons for the statements in the following proof.***

THEOREM OF DESARGUES: If two triangles are such that the lines through their corresponding vertices are concurrent, then the intersections of their corresponding sides are collinear.

*Proof:* Let the two triangles be $ABC$ and $A'B'C'$. Let the lines through $A$ and $A'$, through $B$ and $B'$, and through $C$ and $C'$ meet at $O$.

Let $D$ be the intersection of sides $AB$ and $A'B'$.
Let $E$ be the intersection of sides $BC$ and $B'C'$.
Let $F$ be the intersection of sides $AC$ and $A'C'$.

We must show that $D, E, F$ are collinear. We know that any two distinct points determine a unique line (why?) and that any two lines meet in a unique point (why?); so we can find the following points:

$X$ the intersection of $AB'$ and $CC'$,
$Y$ the intersection of $AC$ and $B'C'$,
$Z$ the intersection of $YO$ and $AB$, and
$W$ the intersection of $YO$ and $A'B'$.

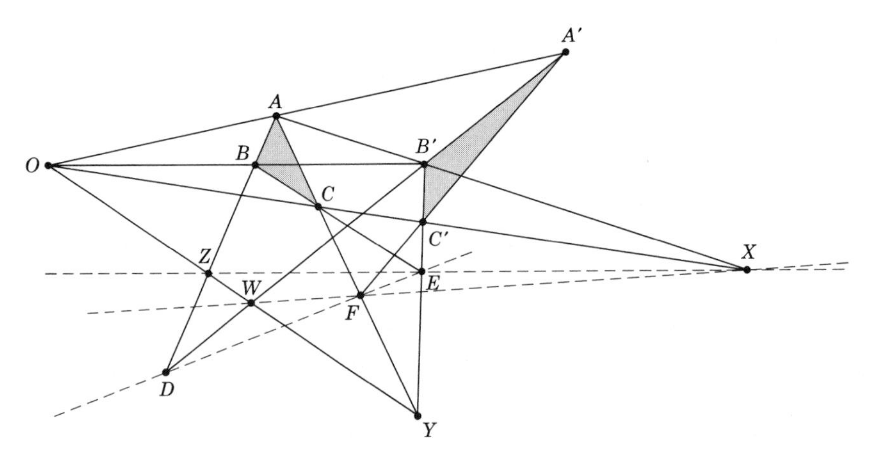

The figure $BCOYB'A$ is a hexagon with alternate vertices, $B$, $O$, $B'$ and $C$, $Y$, $A$ on two lines. Therefore, $E$, $X$, $Z$, the intersections of its opposite sides, are collinear (why?). In a similar way in hexagon $A'C'OYAB'$, $F, X, W$ are collinear (why?). Finally in hexagon $B'YAZXW$ the intersections of $B'Y$ and $XZ$, $YA$ and $WX$, and $AZ$ and $WB'$ are collinear (why?). $B'Y$ and $XZ$ intersect in $E$ because $E$ is on $B'Y$ (why?) and $E,X,Z$ are collinear. $YA$ and $WX$ intersect in $F$ because $F$ is on $YA$ (why?) and $F,X,W$ are collinear. $AZ$ (which is $AB$) and $WB'$ (which is $A'B'$) intersect in D. Therefore, it follows that $D,E,F$ are collinear (why?). ■

## Exercises

1. Draw a different illustration for this proof of Desargues' Theorem.
2. Let $ACPB$ be a quadrilateral in which $F$ is the intersection of $AB$ and $CP$, $E$ is the intersection of $BP$ and $AC$, and $D$ is the intersection of $AP$ and $BC$. If $G$ is

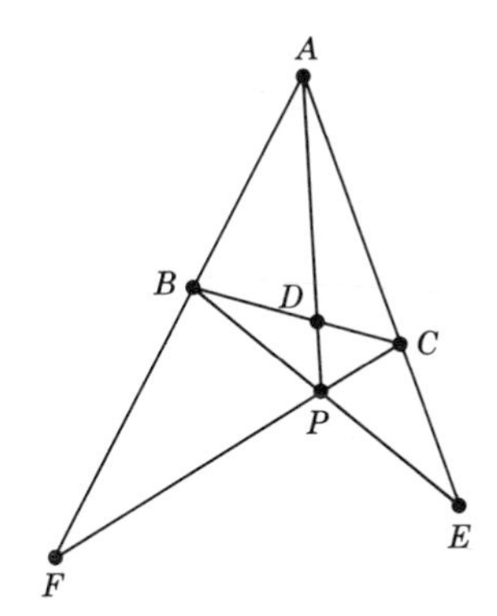


the intersection of $BC$ and $FE$, $H$ the intersection of $CA$ and $FD$, and $K$ the intersection of $AB$ and $DE$, prove that $G,H,K$ are collinear. (Hint: Use Desargues' Theorem.)

3. Prove that in the following figure lines $QP$, $m$, and $n$ meet at a common point.

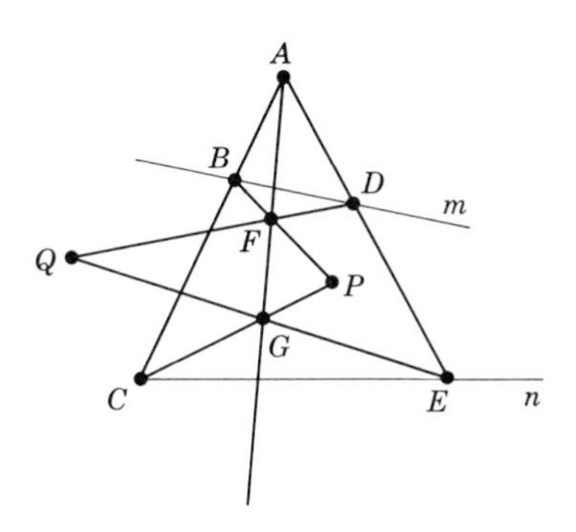

## 6. FINITE PROJECTIVE GEOMETRIES

In the Sections 4 and 5 we talked about an axiom system for projective geometry. We approached the problem much as it happened historically. Much projective geometry existed before the true statements about it were organized into an axiom system, as we explained in Sections 2 and 3.

Recently the process has been reversed. Mathematicians sometimes study an axiom system by itself without reference to any particular objects or relations. If this point of view is adopted, we do not know what the undefined objects and relations represent. The objects simply belong to undetermined sets which have the properties expressed in the axioms. The question of whether a statement is *true* cannot be asked. Instead we focus our attention on whether certain statements can be *validly* proved from the axioms. Those that can be proved are the theorems.

After the axiom system has been studied by itself, we can talk about its *models;* that is, what it represents. The valid statements of the axiom system then become the true statements about each model. It is possible that one axiom system has several different models. For example, in Chapter 2 we could accept the seven properties of a group as the axioms of group theory. Then each of the tables which we found would be a model of group theory. Some of these tables are essentially different than others, so there are different models of group theory.

As further examples of models, let's see if we can find some models for the axioms of projective geometry. To keep the problem simple, let's use as our axiom system A1 to A3. Let's let the points represent beads and the lines represent wires. If an axiom says "a point is on a line," in the model it means "a bead is on a wire."

***Explain why this ornament is not a model of A1 to A3.***

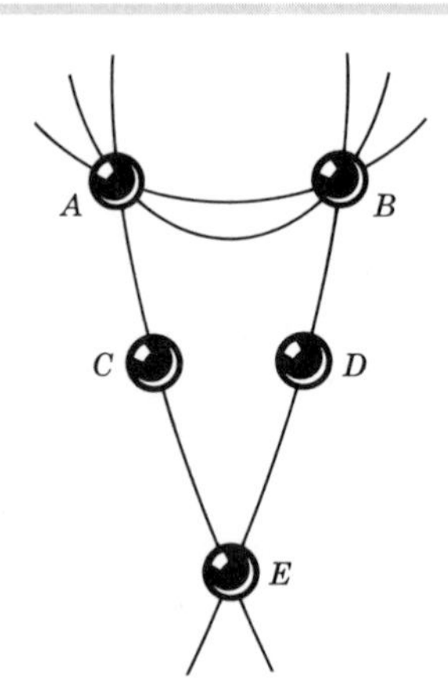

In this ornament the *two wires* through beads *A* and *B* contradict A1. There is no wire through beads *C* and *D*, so again A1 is contradicted. The third axiom is satisfied, however, because of beads *A, B, C,* and *D.* We should also notice that this could not be a model of A1 to A3 because some of the theorems which follow from these axioms are not true: There are only four wires in this ornament (contradicting Theorem 2); two distinct wires go through *both A and B* (contradicting Theorem 3); beads *C, D,* and *E* do not have three wires through them (contradicting Exercise 3 of Section 4); each of the two wires through *A* and *B* does not have three beads on it (contradicting Exercise 4 of Section 4).

***Try to draw a picture of an ornament which is a model of A1 to A3.***

After many trials we might find the following ornament (there are others):

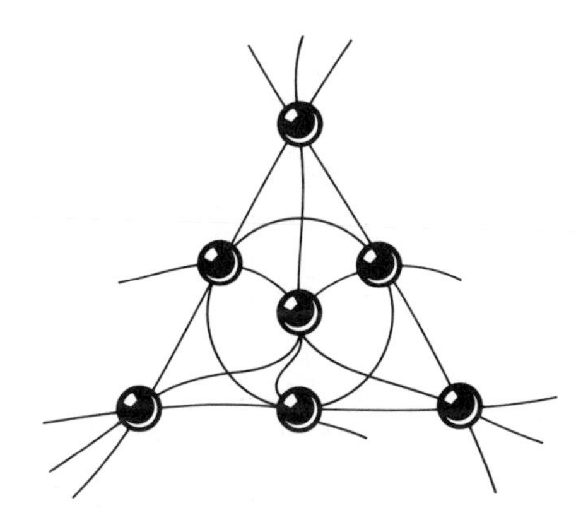

The model consists of seven beads and seven wires. The shape of each wire is not important since the axioms only describe how the wires are related to the beads.

If we want, we can think of the configuration of the ornament as a finite geometry. The points of the axiom system represent points as we usually understand them. The lines of the axiom system represent certain collections of points. If "a point is on a line" in the axioms, then "a point is a member of a line" in this model. The concept of lines in this model is different than our usual concept of lines. These lines have "gaps" in them; the lines are collections of isolated points. The dashed lines in the picture on the next page merely indicate which points are collected together to form lines. This interpretation of the axioms is a second model.

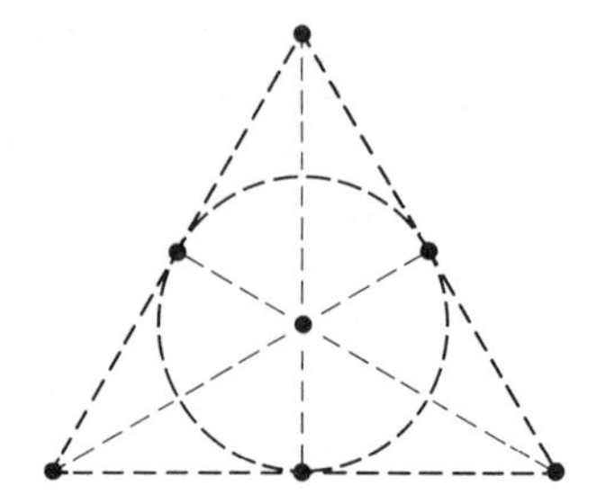

A third model would be to interpret the points as people and the lines as committees. If "a point is on a line" in the axioms, then "a person is on a committee" in this model. A1 to A3 are then true statements about these people and committees.

***Arrange seven people into committees so that A1 to A3 are true.***

If the people are called *A, B, C, D, E, F,* and *G,* the following columns represent the committees. Axioms A1 to A3 are satisfied.

| Committee: | 1 | 2 | 3 | 4 | 5 | 6 | 7 |
|---|---|---|---|---|---|---|---|
| | *A* | *C* | *E* | *A* | *C* | *E* | *F* |
| | *B* | *D* | *F* | *G* | *G* | *G* | *B* |
| | *C* | *E* | *A* | *D* | *F* | *B* | *D* |

We have found three models of A1 to A3: the ornament, the finite geometry, and the committees. We must admit, however, that although these models are about different things, they are essentially the same. The finite geometry could be labelled with *A, B, C, D, E, F,* and *G* so that we could read off the committees. The ornament and the finite geometry are obviously similar.

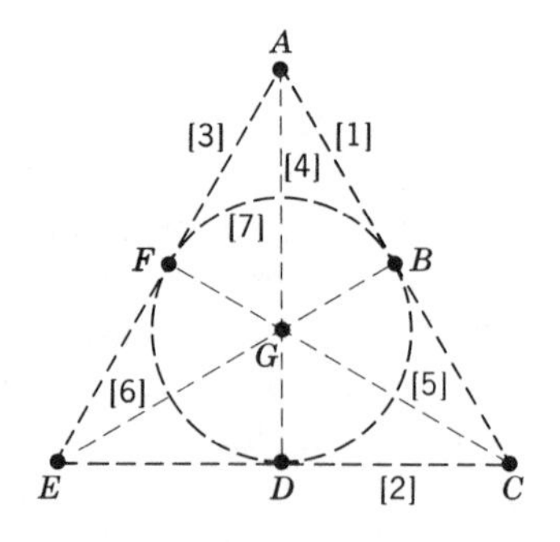

We might wonder if this axiom system has any models which are essentially different. It does. There is a thirteen-point, thirteen-line geometry which satisfies the axioms. If we use the letters *A* to *M* to represent the points, the lines are represented by the columns of the following table. "On" in the axiom means "is an element of" in this model.

| Line: | [1] | [2] | [3] | [4] | [5] | [6] | [7] | [8] | [9] | [10] | [11] | [12] | [13] |
|---|---|---|---|---|---|---|---|---|---|---|---|---|---|
| | *A* | *A* | *A* | *A* | *B* | *B* | *B* | *C* | *C* | *C* | *D* | *D* | *D* |
| | *B* | *E* | *H* | *K* | *E* | *F* | *G* | *E* | *F* | *G* | *E* | *F* | *G* |
| | *C* | *F* | *I* | *L* | *H* | *I* | *J* | *I* | *J* | *H* | *J* | *I* | *H* |
| | *D* | *G* | *J* | *M* | *K* | *L* | *M* | *M* | *K* | *L* | *L* | *M* | *K* |

This table represents a geometry that satisfies A1 to A3. It is essentially different than the seven-point geometry. It seems that we should be speaking of projective *geometries,* rather than projective geometry. There are different projective geometries.

During this century, considerable work has been done on finite projective geometries. Our seven-point model is called a model of order 2. If the *order of a geometry* is *n,* it contains $n^2 + n + 1$ points and the same number of lines. In a geometry of order *n* there are $n + 1$ points on each line and $n + 1$ lines through each point. For order 2, $2^2 + 2 + 1 = 7$ so there are seven points, seven lines, three points on each line, and three lines through each point.

At the present time, some of the models that have been found are for $n = 2$, 3, 4, 5, 7, 8, and 9. It has been proved that there is no model of order 6. At present, the question of whether there is a model of order 10 is unanswered. Here we have another example of an unsolved problem in mathematics. We might feel that, since there are only a finite number of of points ($10^2 + 10 + 1 = 111$) and lines involved in this problem, a high speed computer could simply try all the possibilities. Unfortunately the number of possibilities is so great that even one hundred hours of computer time on a contemporary computer would not begin to exhaust all the cases. Hopefully, in the future, someone will find some general way to approach this problem.

## Exercises

**1.** Explain why an ornament consisting of three beads and three wires in the shape of a triangle is not a model of A1 to A3.

**2.** Show that the thirteen-point table is a model of A1 to A3.

***3.** Draw a picture of an ornament that corresponds to the thirteen-point table.

4. Assume that a projective geometry has $n+1$ points on each line and $n+1$ lines through each point. Prove that the least number of points which could be in the geometry is $n^2+n+1$.
5. Draw a lattice diagram of the seven-bead ornament as follows: Let each bead, each wire, and the entire ornament be an object in the lattice. In addition, add another object called $O$. (This is needed for the diagram to be a lattice. You will see why in Exercise 7.) Draw the lattice diagram so that an object $A$ is below another object $B$ and connected by an ascending line if *$A$ is on $B$* in the ornament and $A \neq B$. The added object $O$ is on every other object.
6. Use the lattice diagram of Exercise 5 and describe the position in the lattice of the smallest object which joins (goes through) two objects. For example, using the letters and numbers of the seven-point geometry, wire [6] joins $E$ and $B$ and is the smallest such object; the entire ornament joins $A$ and wire [2] and is the smallest such object; the join of $O$ and wire [2] is wire [2].
7. Use the lattice diagram of Exercise 5 and describe the position in the lattice of the largest object which is on two objects (the intersection of the two objects). For example, using the letters and numbers of the seven-point geometry, $A$ is the intersection of wire [3] and wire [1]; the intersection of $G$ and the entire ornament is $G$; the intersection of beads $E$ and $D$ is the object $O$ (so $O$ is actually the empty set).

## 7. WHAT HAVE WE LEARNED FROM PROJECTIVE GEOMETRY?

***Try to summarize some of the aspects of mathematics which we have learned by studying projective geometry.***

We have seen again that the first source of a mathematical idea is sometimes a *concrete* situation—the Renaissance technique of painting, or the casting of shadows. The *abstraction* of the real life situation, however, does not completely conform to our intuition. Our intuition tells us that two lines do not always have a point in common. In projective geometry we assume that two lines always do have a point in common.

We have learned that geometry can be studied from different points of view: as the *invariants* under certain changes, or as the *content* of certain statements. This second approach, based on an *axiom system,* also admits of different points of view. We can organize the *true* statements about an already existing geometry, or we can study the geometry abstractly and focus only on the *valid* statements (those which can be proved from the axioms). In the second case we can then look for *models* of the axiom system; there may be many models which may be essentially different from each other.

In working Exercises 5, 6, and 7 in Section 6, we also found the lattice structure appears in finite geometries. *Lattices* also appeared in the study of the divisors of a number, and in the set of normal subgroups of a group. Once again we see a certain unity between seemingly unrelated topics in mathematics. It may be worthwhile to think further about lattices, and we will do this in Chapter 4.

## Exercises

**READ FROM ONE OR MORE OF THE FOLLOWING**

Caillet, Émile, *Pascal, the Emergence of Genius,* Harper and Brothers, New York, 1961.

Dorwart, Harold L., *The Geometry of Incidence,* Prentice-Hall, Englewood Cliffs, New Jersey, 1966, pp. 121–140.

*Kline, Morris, "Projective Geometry," *Scientific American,* Vol. 192, No. 1, January, 1955, pp. 80–86.

Von Helmholtz, Hermann, "On the Origin and Significance of Geometrical Axioms," in *The World of Mathematics,* Vol. 1, J. R. Newman, Simon and Schuster, New York, 1956, pp. 647–666.

The article on Projective Geometry in a recent edition of the *Encyclopaedia Britannica.*

## 8. WHAT IS GEOMETRY?

Projective geometry is only one of many geometries. The following diagram illustrates the connection between some of the main two-dimensional geometries.

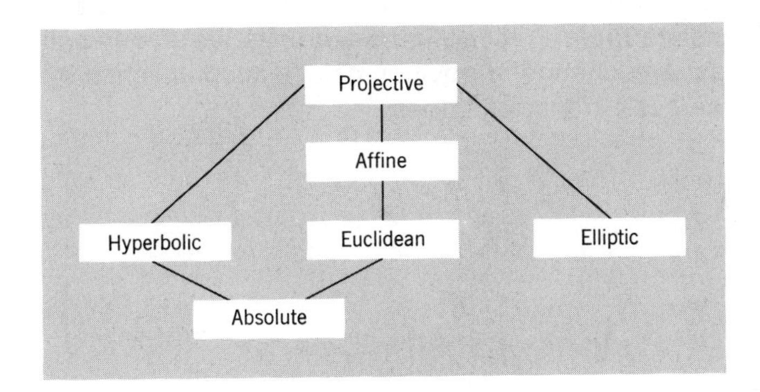

*Projective geometry* appears at the top because it is the most general of these geometries and has the fewest properties. We described projective geometry as the study of the properties of figures preserved by their shadows. We saw that collinearity, concurrency, triangles, and hexagons can be studied in projective geometry.

An *affine geometry* is not quite as general as a projective geometry. An affine geometry has all of the properties of a projective geometry and some additional ones. An affine geometry can be described as the study of the properties of figures preserved by casting shadows from a light which sends parallel rays (something like a fluorescent light). The shadow will be formed by parallel lines.

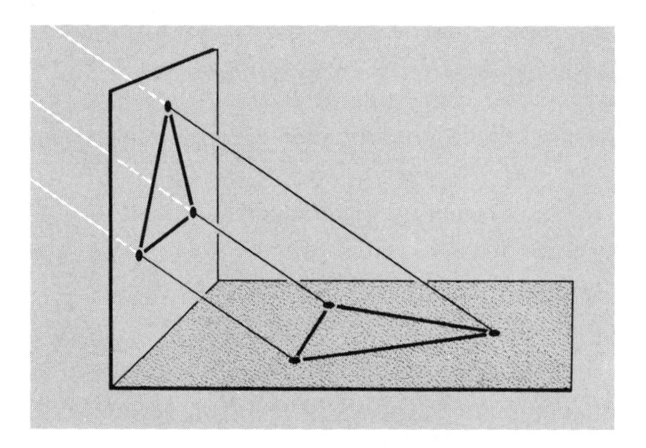

Besides the projective properties, some of the notions we study in affine geometry are betweenness, ratios of distances on a line, parallelism, parallelograms, trapezoids, and ellipses.

Euclidean geometry is less general than projective geometry and affine geometry. It is less general in the sense that we are no longer studying properties preserved by shadows, but the properties preserved by more restricted transformations (changes). Imagine that we have a plane with figures on it and we are allowed to change the position of the figures. We may rotate a figure about a point, move the figures parallel to themselves (translate them), reflect figures in lines, or do any combination of these things. Any change of position can be accomplished by a combination of these basic transformations.

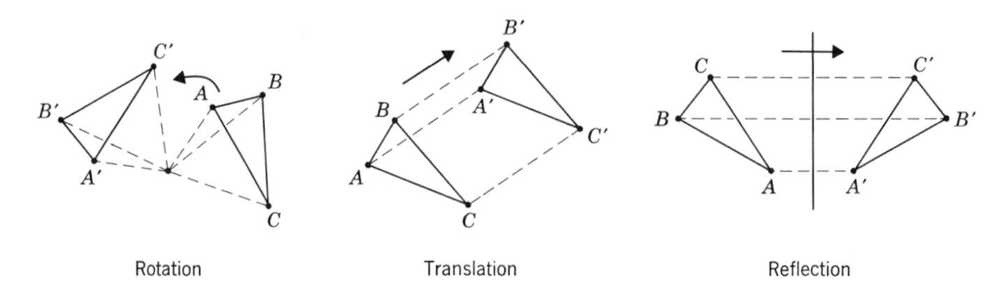


Rotation Translation Reflection

Table 3–1

| Euclidean | Hyperbolic | Elliptic |
|---|---|---|
| **The sum of the angles of a triangle is 180°.** | **The sum of the angles of a triangle is less than 180°.** | **The sum of the angles of a triangle is more than 180°.** |
| **Given a point not on a line there is one and only one line through the point and parallel to the given line.** | **Given a point not on a line there are an infinite number of lines through the point which do not intersect the given line.** | **There are no parallels.** |
| **Squares exist.** | **There are no squares.** | **There are no squares.** |
| 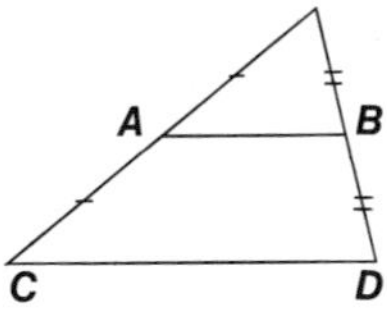 A, B, C, D $AB = (1/2)\ CD$ | $AB < (1/2)\ CD$ | $AB > (1/2)\ CD$ |
| 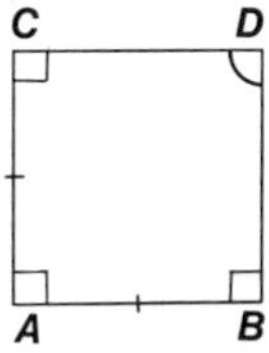 C, D, A, B $D = 90°$ | $D < 90°$ | $D > 90°$ |
| 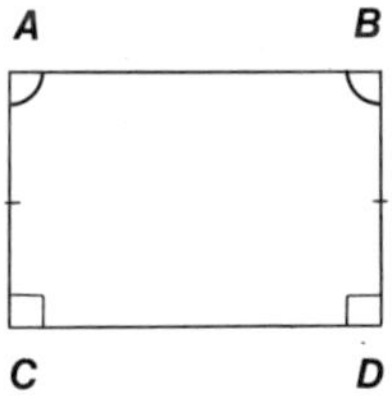 A, B, C, D $A = B = 90°$ | $A = B$<br>$B < 90°$ | $A = B$<br>$B > 90°$ |
| **Two similar triangles need not be congruent.** | **If two triangles are similar, then they are congruent.** | **If two triangles are similar, then they are congruent.** |
| **Lines are infinitely long.** | **Lines are infinitely long.** | **Lines are finite in length.** |

*Euclidean geometry* is the study of the properties of figures preserved under changes of position. Distance, size of angle, perpendicularity, congruence, similarity, squares, and circles are some of the Euclidean variants. Of course, all of the things studied in affine and projective geometry can also be studied in Euclidean geometry.

The other three geometries in our diagram could also be described as the study of the properties of figures preserved under certain transformations. However, in our short discussion it will be clearer if we use our second approach to geometry and talk about some of the theorems (true statements) about these geometries.

*Hyperbolic* and *elliptic* geometries are known as *non-Euclidean* geometries. Table 3–1 indicates why they are called non-Euclidean.

In spite of the differences between Euclidean and hyperbolic geometries, there are still many statements which are true in both geometries. The collection of all these statements common to the two geometries is called *absolute geometry.*

Our list of geometries is far from complete. There are other two-dimensional geometries as well as many three-, four-, and higher-dimensional geometries. There are even geometries of no dimension at all. Some of these geometries are Euclidean, others are not.

No one would question the value of studying two- and three-dimensional Euclidean geometry. Applications of these geometries surround us—architecture, engineering, design, and surveying, to mention a few things, would be impossible without Euclidean geometry. Elliptic geometry, although it is not Euclidean, does have as a model a situation in the real world. The geometry on the surface of a sphere is elliptic. Since the earth approximates a sphere, elliptic geometry is useful in navigation and aviation.

Geometries which do not seem to have the properties of the real world were developed in the 19th century. Their lack of conformity with the intuitive idea of geometry had much to do with a change of attitude toward axiom systems. Axioms were originally supposed to be self-evident statements about the real world. Since geometries like hyperbolic did not seem to have the real world as a model, the emphasis on the axiom system being *true* in the real world was replaced by an interest in the *validity* of the proof of the statement from the axioms. The age of pure mathematics was beginning.

Pure mathematics, mathematics pursued as an art and for its own sake, has always had opponents. What possible good can come from a geometry which does not resemble the real world? This seems to make mathematics nothing more than an ingenious game played by highly skilled and rigorously logical contestants. One mistake in this line of thought is the presumption that these geometries do not reflect any part of the real world. It now appears that the geometry of subatomic particles moving at extremely high velocities, and the geometry of the extremely large distances involved in theories of the universe, may not be Euclidean. A non-Euclidean geometry of four dimensions was used by Einstein for his relativistic model of the universe. Since then, a variety of different models of the universe have been proposed using various geometries. As long as the geometry of cosmic space is open to question, investigations in any geometry are highly practical.

Unfortunately, there is no way of predicting ahead of time which aspects of pure mathematics will eventually have applications. Since this is the case, we must be careful about labelling some part of mathematics as trivial because it does not seem to be practical.

## Exercises

**READ AT LEAST ONE OF THE FOLLOWING**


Bergamini, David, and the Editors of Life, *Mathematics,* Life Science Library, Time Incorporated, New York, 1963, pp. 149–167.

Coxeter, H. M. S., "Non-Euclidean Geometry," *The Mathematical Sciences,* Edited by COSRIMS, M.I.T. Press, Cambridge, Mass., 1969, pp. 52–59.

*Einstein, Albert, "On the Generalized Theory of Gravitation," *Scientific American,* Vol. 182, No. 4, April, 1950, pp. 13–17.

*Gamow, George, "Gravity," *Scientific American,* Vol. 204, No. 3, March, 1961, pp. 94–106.

*Hahn, Hans, "Geometry and Intuition," *Scientific American,* Vol. 190, No. 4, April, 1954, pp. 84–91.

*Kline, Morris, "Geometry," *Scientific American,* Vol. 211, No. 3, September, 1964, pp. 60–69.

*LeCorbeiller, P., "The Curvature of Space," *Scientific American,* Vol. 191, No. 5, November, 1954, pp. 80–86.

Panofsky, Erwin, "Dürer as a Mathematician," in *The World of Mathematics,* Vol. 1, J. R. Newman, Simon and Schuster, New York, 1956, pp. 603–621.

Weyl, Hermann, "Symmetry," in *The World of Mathematics,* Vol. 1, J. R. Newman, Simon and Schuster, New York, 1956, pp. 671–724.

The article on Geometry, Hyperbolic Geometry, Euclidean Geometry, or Elliptic Geometry in a recent edition of the *Encyclopaedia Britannica.*


## Test

**1.** Find the shadow of Point $P$.

$l$

$O$

$P$

$m$

2. Find the shadow of triangle $ABC$.

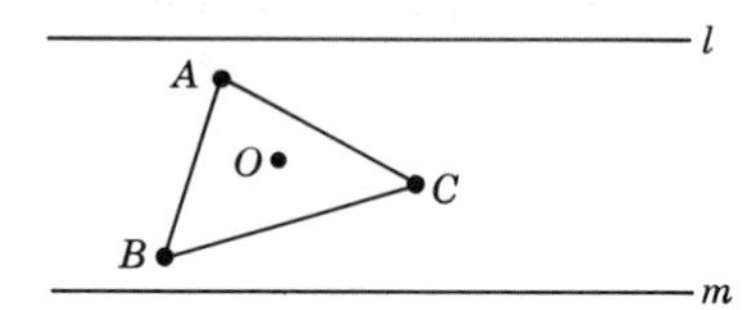


3. State Desargues' Theorem.
4. Draw a figure to illustrate Pappus' Theorem.
5. Which of the following are projective invariants?

| | | |
|---|---|---|
| circle | perpendicularity | distance |
| hexagon | parallelism | concurrency |
| square | collinearity | trapezoid |
| | | a point being on a line |

6. Explain the relationship between projective, affine, and Euclidean geometries.
7. State three ways in which hyperbolic geometry is different from Euclidean geometry.
8. In this figure use the Theorem of Desargues to prove that $R, S, T$ are collinear.

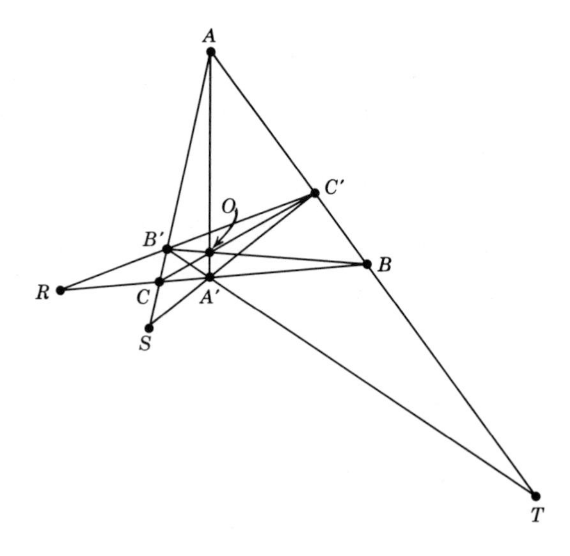


9. Use Pappus' Theorem and a straight edge to find a point on the left side of the lake and on line $l$ without drawing any line across the lake. (Hint: Let $l$ be

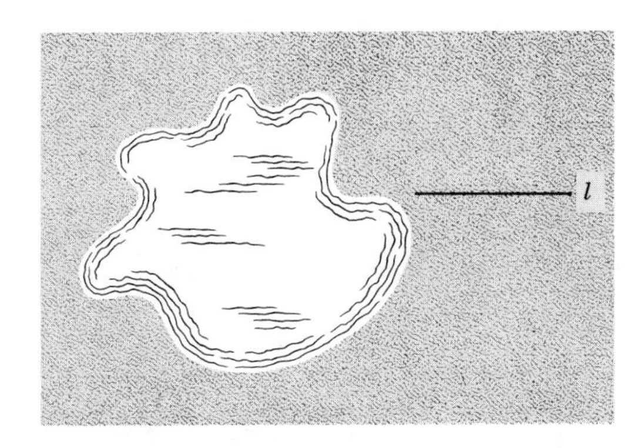


the line on which the opposite sides of a hexagon satisfying the Theorem of Pappus intersect.)

**10.** Summarize the outside reading you have done about geometry.

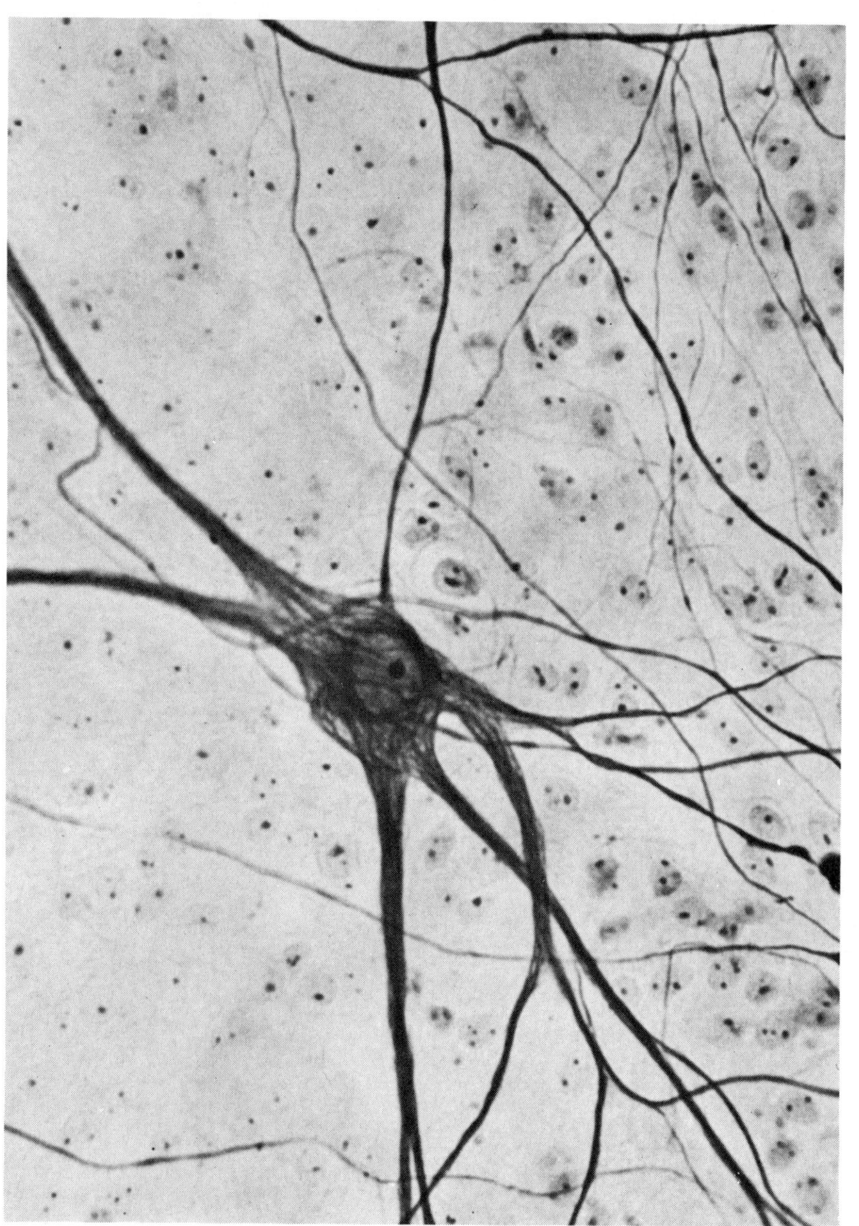

Eric V. Grave

4

# FOUNDATIONS

## Looking at Many Things at Once

**Boolean algebra, a special type of lattice, is currently being used to analyze the nervous system.**

Three times already we have seen a generalized type of order which we called a lattice. The divisors of a natural number ordered by *divides* form a lattice (Chapter 1). A lattice is formed by the normal subgroups of a group if they are ordered by *is a subgroup of* (Chapter 2). The elements of a finite projective geometry ordered by *is on* also form a lattice (Chapter 3). In Section 1 of this chapter we will look at a fourth area in which lattices appear. The remainder of the chapter will be devoted to a precise description of a lattice and a study of some of its properties.

***Reread Section 6 of Chapters 1, 2, and 3 before reading further.***

## 1. THE LATTICE OF SUBSETS OF A SET

A travel agency offers a tour of three major European cities—London, Paris, and Rome. Besides offering this entire tour, it also arranges smaller tours, such as a trip to London and Rome, or one to Paris and Rome, or even a tour of just Rome. Let's represent the largest tour by $\{L,P,R\}$. The smaller ones could be represented in a similar way, as $\{P,R\}$ for the tour of Paris and Rome, and $\{R\}$ for the trip to just Rome. We could call one tour a subtour of a second, if it is a part of the second tour; that is, if the cities on the first tour are on the second tour.

***List all the subtours of*** $\{\mathbf{L},\mathbf{P},\mathbf{R}\}$.

There are eight possible subtours. The ones we think of first are $\{L,P\}$, $\{L,R\}$, $\{P,R\}$, $\{L\}$, $\{P\}$, and $\{R\}$. In addition we could call $\{L,P,R\}$ a subtour of itself, and $\{\quad\}$, the tour of no cities, could also be called a subtour of $\{L,P,R\}$.

If we look at any two of these eight tours, it is possible that one of them is a subtour of the other; but this need not be true. The tour $\{R\}$ is a subtour of $\{L,R\}$, but $\{P,R\}$ is not a subtour of $\{L,R\}$. We can summarize by saying that *is a subtour of* is a relation on these eight tours.

***Draw a diagram which indicates the relation* is a subtour of *for all the subtours of* {L,P,R}.**

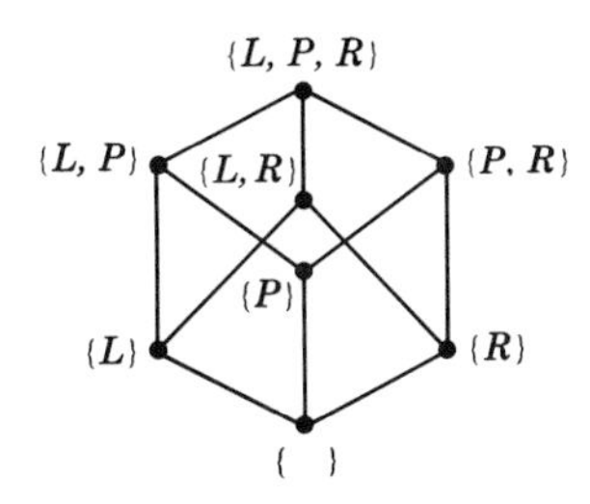


This diagram indicates which tours are subtours of others. One tour is a subtour of another if it is below the other and connected to it by an ascending line; any tour, of course, is a subtour of itself. The diagram is a *lattice* and we can say that the tours are *ordered* by *is a subtour of.* Diagrams similar to this one are the subject of this chapter.

In the example of the lattice of subtours of a tour, the fact that we are talking specifically about tours has little to do with the lattice structure. The structure is there because one tour can be a part of another tour. We can formulate a more general example of a lattice if we forget about tours and simply speak of a *set* of objects $\{L,P,R\}$ and all of its *subsets,* $\{L,P,R\}$, $\{L,P\}$, $\{L,R\}$, $\{P,R\}$, $\{L\}$, $\{P\}$, $\{R\}$, and $\{\ \ \}$. A set **X** *is a subset of* a set **Y**, and we write $\mathbf{X} \subseteq \mathbf{Y}$, if every element of **X** is an element of **Y**. The subsets of $\{L,P,R\}$ are ordered by the relation $\subseteq$, *is a subset of.*

***Find all the subsets of* {A,B} *and form the lattice of these subsets under* $\subseteq$.**

The subsets of $\{A,B\}$ are $\{A,B\}$, $\{A\}$, $\{B\}$, and $\{\ \ \}$. The lattice is:

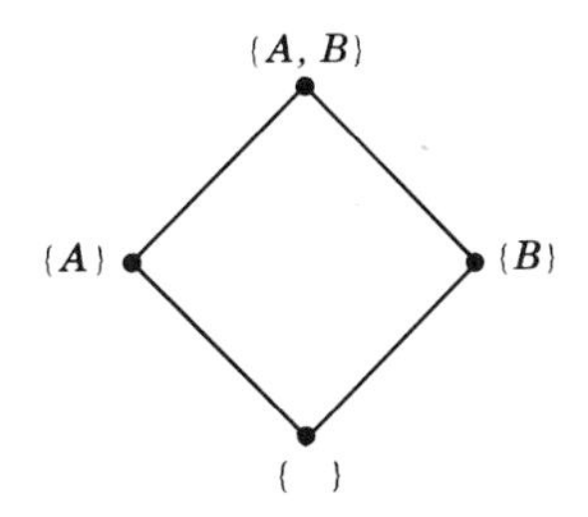

***Draw the lattice of all of the subsets of* {K, L, M, N} *ordered by the relation* ⊆.**

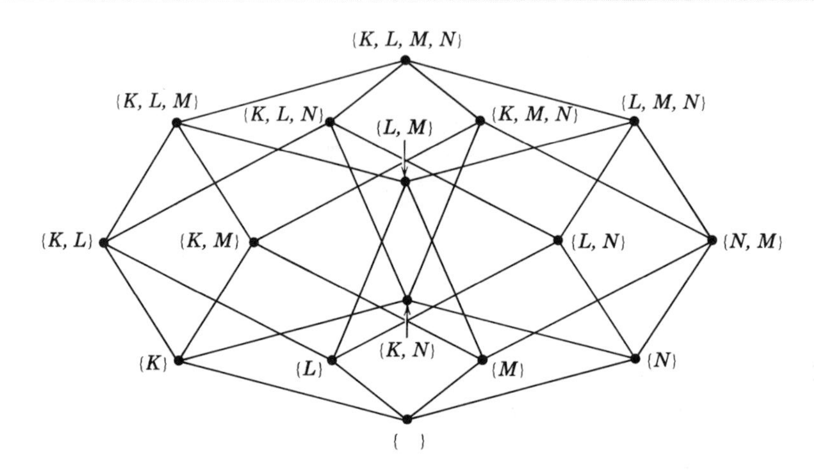


When we speak of the subsets of $\{K, L, M, N\}$ it is possible to consider, in addition to the relation ⊆, operations on these subsets. One such operation is called *union*, and written ∪; it assigns to two sets the unique set which consists of all of the elements in either or both of the subsets. For example,

$$\{K, L, M\} \cup \{L, N\} = \{K, L, M, N\},$$
$$\{K, L\} \cup \{N\} = \{K, L, N\},$$
$$\{K, L\} \cup \{\ \ \} = \{K, L\}, \text{ and}$$
$$\{\ \ \} \cup \{\ \ \} = \{\ \ \}.$$

Another operation on the subsets is called *intersection* and is written ∩. The intersection of two sets is the set of all elements common to these two sets. For example,

$$\{K, L, M\} \cap \{L, N\} = \{L\},$$
$$\{K, L\} \cap \{N\} = \{\ \ \},$$
$$\{K, L\} \cap \{K, L\} = \{K, L\}, \text{ and}$$
$$\{K, L\} \cap \{\ \ \} = \{\ \ \}.$$

We might wonder if there is any relation between these operations of ∪ and ∩ on the subsets of $\{K, L, M, N\}$ and the lattice of subsets under ⊆.

***In the lattice of subsets of* {K, L, M, N}*, describe the position of the subset which is the union of two subsets.***

If two subsets **A** and **B** are connected by an ascending line, then $\mathbf{A} \cup \mathbf{B}$ is the higher of **A** and **B**; if $\mathbf{A} \neq \mathbf{B}$ and **A** and **B** are not connected by an ascending line, then $\mathbf{A} \cup \mathbf{B}$ is the closest subset above both of them and connected to them by lines. If $\mathbf{A} = \mathbf{B}$, then $\mathbf{A} \cup \mathbf{B} = \mathbf{A}$.

***In the lattice of subsets of $\{\mathbf{K}, \mathbf{L}, \mathbf{M}, \mathbf{N}\}$ describe the position of the subset which is the intersection of two subsets.***

If two subsets **A** and **B** are connected by an ascending line, then $\mathbf{A} \cap \mathbf{B}$ is the lower of **A** and **B**; if $\mathbf{A} \neq \mathbf{B}$ and **A** and **B** are not connected by an ascending line, then $\mathbf{A} \cap \mathbf{B}$ is the closest subset below both of them and connected to them by ascending lines; if $\mathbf{A} = \mathbf{B}$, then $\mathbf{A} \cap \mathbf{B} = \mathbf{A}$.

The order relation in a lattice seems to be closely connected to two operations on the elements. (This was also implied in some of the exercises of Section 6 in the first three chapters). The operations assign an element to each pair of elements. As one operation makes this assignment it moves up in the lattice, let's call this operation *cup.* Let's call the other operation, which moves down as it makes the assignments, *cap.* In the four types of lattices we have seen that the order relation and two operations are as follows:

| Chapter | Elements | Order Relation | Cup | Cap |
|---|---|---|---|---|
| 1 | All divisors of a number | $\mid$ | lcm | gcd |
| 2 | All normal subgroups of a group | is a subgroup of | "product" | intersection |
| 3 | All elements of a finite geometry | is on | join | intersection |
| 4 | All subsets of a set | $\subseteq$ | $\cup$ | $\cap$ |

***In order to be sure you understand this table, work again Exercises 2 and 3 of Section 6, Chapter 1; Exercises 6 and 7 of Section 6, Chapter 2; and Exercises 5, 6, and 7 of Section 6, Chapter 3.***

Since lattices appear in so many different parts of mathematics, it would be economical to study them abstractly and then apply the results to the various systems which form lattices. In Sections 2, 3, and 4, we will find an axiom system for lattices, prove a theorem about lattices, and apply this theorem to the four examples of lattices which we have already seen.

## Exercises

**1.** In the set of all subsets of $\{A,B,C,D,E\}$ are the following true or false?
(a) $A \subseteq \{A,B\}$ (d) $\{A,B\} \subseteq \{B\}$ (g) $\{C,D\} \subseteq \{D,E,F\}$
(b) $\{B\} \subseteq \{A,B\}$ (e) $\{\ \} \subseteq \{A\}$ (h) $\{A,E\} \subseteq \{A,C,E\}$
(c) $\{A,B\} \subseteq \{A,B\}$ (f) $\{C\} \subseteq \{A,B\}$ (i) $\{A,B,C,D\} \subseteq \{A,B,C,D\}$

**2.** In the set of all subsets of $\{A,B,C,D,E\}$ find:
(a) $\{A,B\} \cup \{A,C\}$ (d) $\{D\} \cup \{E\}$ (g) $\{A,B\} \cap \{B,C\}$
(b) $\{A\} \cap \{C,D\}$ (e) $\{C,D\} \cup \{\ \}$ (h) $\{A,B\} \cup \{A,B,C\}$
(c) $\{A\} \cap \{A,B\}$ (f) $\{\ \} \cap \{A\}$ (i) $\{A,B\} \cap \{A,B,C,D\}$

***3.** Draw a diagram of the lattice of all subsets of $\{A,B,C,D,E\}$.

**4.** Consider the lattice of all the divisors of 12 ordered by $|$. Find lcm(2,3), lcm(2,1), lcm(12,3), lcm(4,6), and lcm(3,3) by looking at the lattice. Find gcd(2,3), gcd(2,1), gcd(12,3), gcd(4,6), and gcd(3,3) by looking at the lattice.

**5.** Consider the lattice of normal subgroups found in Exercise 4 in Section 6, Chapter 2. Let $\mathbf{H} = \{A\}$, $\mathbf{K} = \{A,D\}$, $\mathbf{L} = \{A,C,E\}$, and $\mathbf{M} = \{A,B,C,D,E,F\}$. Find the intersection of **L** and **K**, **H** and **L**, **K** and **M**, and **M** and **M** by looking at the lattice. Find the "product" of each of these same pairs by looking at the lattice.

**6.** In the lattice of the seven-point geometry (seven-bead ornament) found in Exercise 5 of Section 6, Chapter 3, find the intersection of $A$ and $C$, $A$ and [4], $A$ and [5], [4] and [2], [4] and [5], the whole ornament and $E$, and $E$ and $O$ by looking at the lattice. Find the joins of the same pairs by looking at the lattice.

### 2. AXIOMS FOR LATTICE THEORY

We have already seen several examples of orderings which we called lattices; however, we have never described precisely the nature of a lattice. This may seem a little backward—calling something a lattice before we know precisely what a lattice is. Actually, this is how mathematics develops. Mathematicians begin to notice similarities among various examples. The examples are studied individually first, then someone compiles a precise list of the common properties of the examples. These common properties form an abstract theory, and the individual examples are models of the theory (see Section 6, Chapter 3).

We now want to study lattice theory abstractly. This means that we want to find a characterization of a system which has among its models all the particular lattices we have seen.

***Collect all the lattice diagrams that you have seen so far.***

Some of the diagrams are given in Fig. 4–1 on the next page.

In a general theory of lattices we want to study the characteristics shared by all the preceding examples. The examples are concerned with various objects—numbers, subgroups, points, lines, sets, and so on. The common characteristic in each example is that it is concerned with *objects,* so we will begin our development of lattice theory by assuming that we have a set of objects, $\{A,B,C,\ldots\}$. (We are not implying by this notation that there are three or more objects, but we do assume that there is at least one object.)

Besides being concerned with objects, each lattice diagram indicates that certain objects are related to each other. In the particular examples, the relation could be $A|B$, *A is a subgroup of B*, *A is on B*, or $A \subseteq B$. The common characteristic noticed here is that certain objects are related to each other and other objects are not. In each lattice diagram in the examples, $A$ is related to $B$ if $A$ is below $B$ and connected by an ascending line or if $A = B$. Perhaps we could indicate whether or not two objects are related by the notion of "below." Unfortunately "below" is not a mathematical concept—it is a physical notion. We will have to find a way to describe which objects are related to each other and which are not without using the physical description. We will now see how this can be done.

Along with the assumption that we have a set of objects $\{A,B,C,\ldots\}$ we will assume that there is a relation $\preceq$ on certain pairs of these objects considered in a certain order. At this point we do not know the nature of $\preceq$. It will be described a little later by the axioms. We will read $A \preceq B$ as "$A$ is below or equal $B$" and $A \npreceq B$ as "$A$ is not below and not equal $B$." We will use the word "below" to help our intuition, however, we must be very careful in what follows to use this intuition only for inspiration. When it comes to proving something about lattices in general we can use only the axioms which follow, and not the intuitive picture we have of a lattice.

We are now ready to state the axioms for lattice theory. Using only the notion of a set $\{A,B,C,\ldots\}$ with a relation $\preceq$, we want to list some axioms

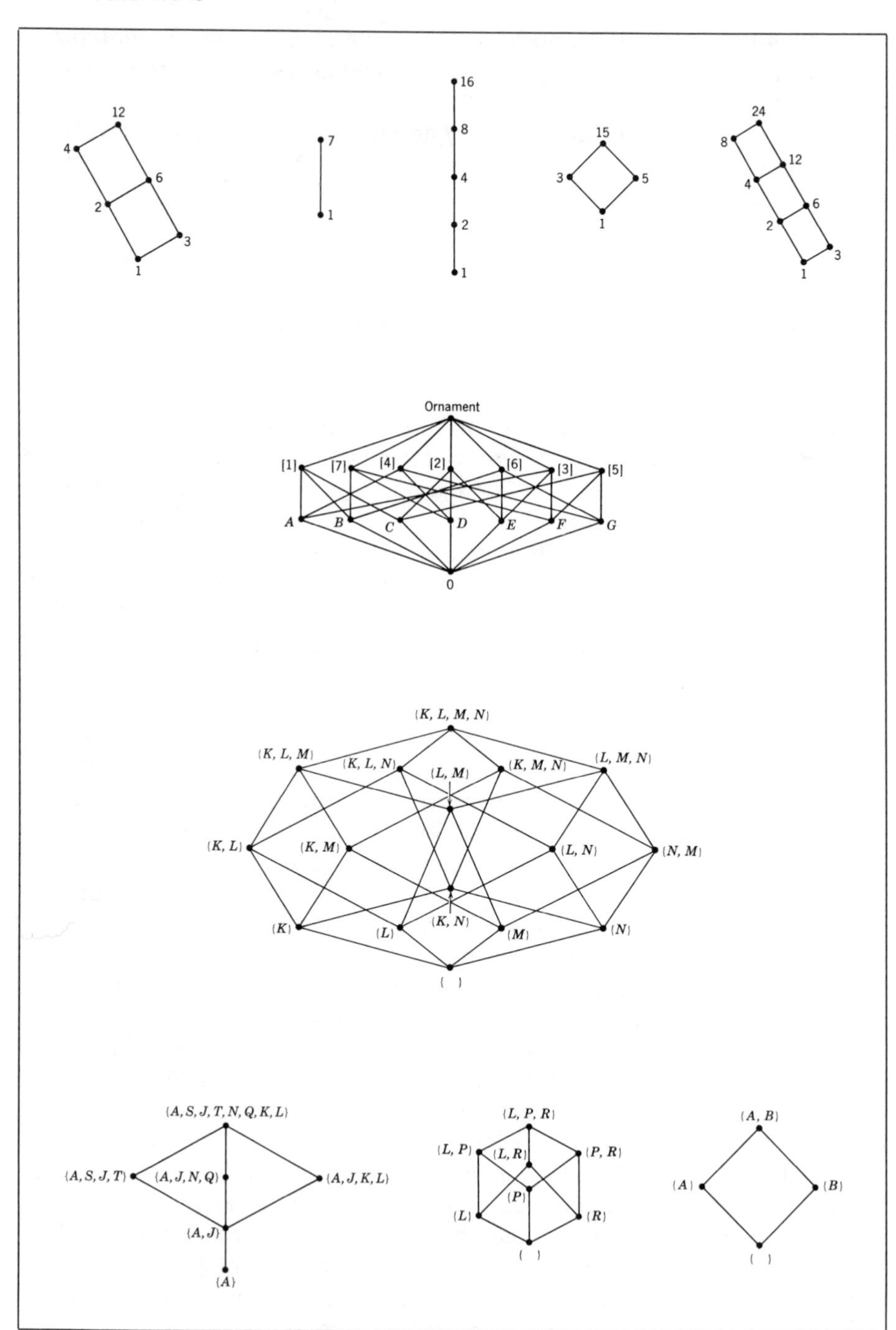

12
4
6
2
3
1
7
1
16
8
4
2
1
15
3
5
1
24
8
12
4
6
2
3
1
Ornament
[1]
[7]
[4]
[2]
[6]
[3]
[5]
A
B
C
D
E
F
G
0
{K, L, M, N}
{K, L, M}
{K, L, N}
{L, M}
{K, M, N}
{L, M, N}
{K, L}
{K, M}
{L, N}
{N, M}
{K}
{L}
{K, N}
{M}
{N}
{ }
{A, S, J, T, N, Q, K, L}
{A, S, J, T}
{A, J, N, Q}
{A, J, K, L}
{A, J}
{A}
{L, P, R}
{L, P}
{L, R}
{P, R}
{P}
{L}
{R}
{ }
{A, B}
{A}
{B}
{ }


**Fig. 4–1.**

(statements) which describe all of the examples we have seen. We might, for example, make the following statement:

*If $A \preceq B$, then $B \npreceq A$.*

---

***Is this statement true about all the lattices you have seen?***

---

The statement is not true because there are counterexamples. For example $5|5$, so $A$ and $B$ could both be 5, then $A \preceq B$ and $B \preceq A$.

---

***Modify the statement so that it is true for all the lattices you have seen.***

---

*If $A \neq B$ and $A \preceq B$, then $B \npreceq A$* would be a correct modification. We could adopt this as one of our axioms.

---

***Can you formulate another true statement about the lattices we have seen? (Hint: Consider the counterexample just given.)***

---

The counterexample suggests that another possible axiom would be: *For all objects $A$, $A \preceq A$.*

---

***In all the examples of lattices we have seen an ascending line can be made up of segments. Formulate an axiom using only our technical terms which expresses this.***

---

The simplest way to express this is : *If $A \preceq B$ and $B \preceq C$, then $A \preceq C$.* We will adopt the three statements just discussed as axioms:

L1. If $A \neq B$ and $A \preceq B$, then $B \npreceq A$.
L2. For all $A$, $A \preceq A$.
L3. If $A \preceq B$ and $B \preceq C$, then $A \preceq C$.

---

***In what ways are the following diagrams different from the lattice diagrams we have seen? Do these diagrams satisfy L1 to L3?***

---

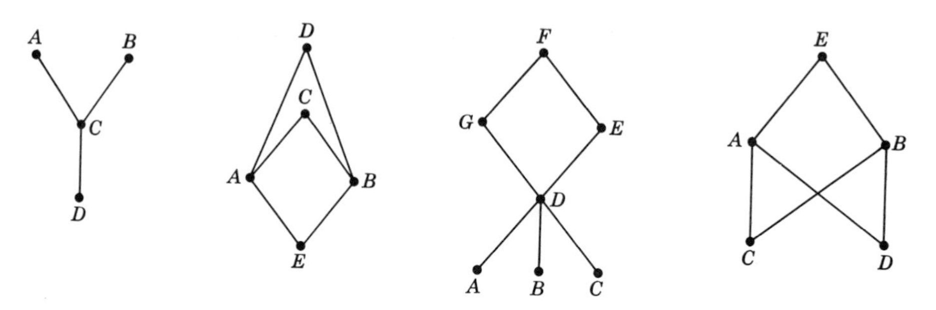


The first diagram is unlike our examples because there is no object above and connected to $A$ and $B$. The second diagram is different from the examples we have seen previously because both $D$ and $C$ are *directly* above and connected to $A$ and $B$ (and $A$ and $B$ are both directly below and connected to $D$ and $C$). The third diagram is not like the examples because $A$ and $B$, $B$ and $C$, and $A$ and $C$ have no object below and connected to them. The last diagram is not like the examples we have seen because $C$ and $D$ are *both directly* below and connected to $A$ and $B$ (and $A$ and $B$ are both directly above and connected to $C$ and $D$).

Historically, as lattice theory was developing it was decided that diagrams similar to these last four are *not* lattice diagrams. Since each of the diagrams satisfies L1 to L3, it should be clear that our list of axioms is not adequate to characterize a lattice.

***Formulate some further axioms which will eliminate the possibility that the last four diagrams are models of the axiom system.***

The reason that the first diagram is not a lattice is that there is no object $X$ such that $A \preceq X$ and $B \preceq X$. In order to make sure that this diagram is not considered to be a lattice diagram we can insist that the following axiom be satisfied.

L4. For all $A, B$, there exists an object $X$ such that $A \preceq X$ and $B \preceq X$.

The second diagram differs from lattice diagrams in that $A \preceq C$, $B \preceq C$, $A \preceq D$, $B \preceq D$, and $C$ and $D$ are both directly above $A$ and $B$. Lattice diagrams have a smallest element among all those elements above $A$ and $B$. For example, in the lattice of divisors of 12, $2|6$, $3|6$, $2|12$, $3|12$, *and* $6|12$. The following axiom eliminates the second diagram from the lattice diagrams.

L5. For all $A, B$, and all $X$ such that $A \preceq X$ and $B \preceq X$, there exists an object $Y$ such that $A \preceq Y$, $B \preceq Y$, and $Y \preceq X$.

Intuitively, L5 means that if several objects are above both $A$ and $B$, then one of these several objects is below all the others.

The third diagram presents the same difficulty as the first diagram, except that the object is missing below $A$ and $B$, $B$ and $C$, and $A$ and $C$. The fourth diagram has a characteristic similar to the second diagram, except that the two objects are directly below $A$ and $B$. These third and fourth diagrams can be eliminated from the lattice diagrams by the following two axioms, respectively. Notice that these axioms are similar to L4 and L5.

L6. For all $A, B$, there exists an object $X$ such that $X \preceq A$ and $X \preceq B$.

L7. For all $A, B$, and all $X$ such that $X \preceq A$ and $X \preceq B$, there exists an object $Y$ such that $Y \preceq A$, $Y \preceq B$, and $X \preceq Y$.

We will define a lattice to be a set of objects $\{A, B, C, \ldots\}$ with a relation satisfying L1 to L7.

**Which of the following are lattices, that is, which diagrams satisfy L1 to L7?**

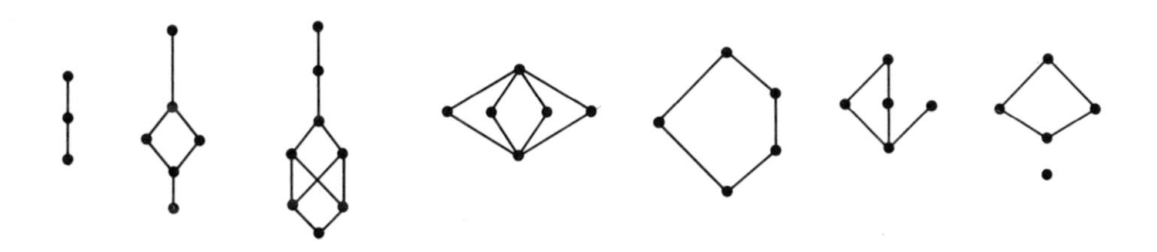

The first, second, fourth, and fifth diagrams represent lattices.

## Exercises

1. A *poset* (partially ordered set) is a set $\{A, B, C, \ldots\}$ with a relation $\preceq$ satisfying L1 to L3. Draw diagrams of all possible three-element posets.
2. Draw all possible four-element posets (see Exercise 1).
3. Which of the posets in Exercise 1 are lattices?
4. Which of the posets in Exercise 2 are lattices?

**5.** Draw five different five-element lattices.

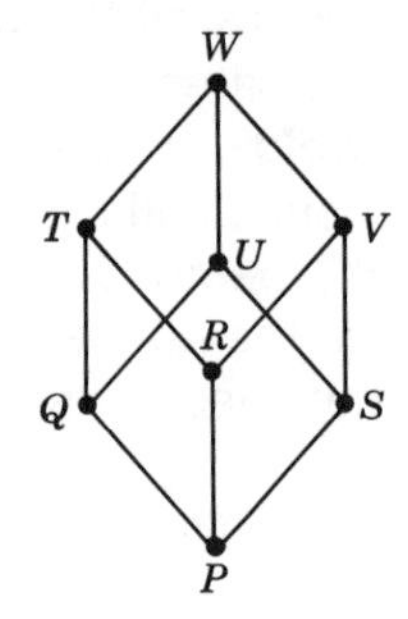


**6.** In this lattice which objects could be $X$ in L4 for $T$ and $U$? For $Q$ and $P$? For $V$ and $V$? For $Q$ and $S$? Use the same pairs and decide which object could be $Y$ in L5.

**7.** In the lattice of Exercise 6 which objects could be $X$ in L6 for $T$ and $U$? For $Q$ and $P$? For $V$ and $V$? For $Q$ and $S$? Use the same pairs and decide which object could be $Y$ in L7.

**8.** Show that the set $\{1, 2, 3, 4, 5\}$ with the relation $\leq$ (less than or equal to) is a lattice. Draw a diagram of this lattice. Would the same set with the relation $\geq$ (greater than or equal to) be a lattice?

## 3. A THEOREM ABOUT LATTICES

In the last section we adopted L1 to L7 as a set of axioms which characterize lattice theory. We should notice, however, that these axioms make no explicit mention of the operation of cup and cap which we found in the examples of lattices in Section 1.

***Which axioms provide a way to define cup?***

Axiom L4 intuitively means that for each pair of objects $A, B$ there is an object $X$ above both of them and connected to them. Axiom L5 guarantees that among such $X$ there is a smallest one $Y$ (that is $Y \preceq X$). Because of these two axioms we can define:

> DEFINITION 1: $A$ cup $B = C$ if and only if $A \preceq C$, $B \preceq C$ and if $A \preceq X$, $B \preceq X$, then $C \preceq X$.

Intuitively, $C = A$ cup $B$ is the smallest object above and connected to both $A$ and $B$.

***Formulate a similar definition of* A cap B*, so that intuitively* A cap B *is the largest object below and connected to* A *and* B.**

DEFINITION 2: $A$ cap $B = D$ if and only if $D \preceq A$, $D \preceq B$ and if $X \preceq A$, $X \preceq B$, then $X \preceq D$.

Notice that by formulating these abstract definitions we have freed cup and cap from the physical notions of up and down that we found necessary to describe our lattice before our axiomatic development.

We might wonder if the operations of cup and cap have some properties similar to those of the operations of addition and multiplication.

***Use the lattice in Exercise 6 of the previous section and experiment to find some properties of cup and cap.***

We might discover some of the following properties.

1. $A$ cup $A = A$, and $A$ cap $A = A$.
2. $A$ cup $B = B$ cup $A$, and $A$ cap $B = B$ cap $A$.
3. $A$ cup ($B$ cup $C$) = ($A$ cup $B$) cup $C$, and $A$ cap ($B$ cap $C$) = ($A$ cap $B$) cap $C$.

***In the lattice of Exercise 6 of the previous section, find two examples of each of the properties just listed.***

These examples, of course, do not prove that Properties 1 to 3 always hold in a lattice. We are looking at only one lattice and at only a few examples in this one lattice. A proof must show that the properties are true for every lattice and for all elements in these lattices. A proof would consist in deducing these properties from the Axioms L1 to L7 of lattice theory. We will prove one of these properties to illustrate proofs in lattice theory.

THEOREM: *$A$ cup ($B$ cup $C$) = ($A$ cup $B$) cup $C$.*

*Proof:* Let $A$ cup ($B$ cup $C$) = $S$, and $B$ cup $C = T$. From Definition 1, $A \preceq S$, $B$ cup $C = T \preceq S$, $B \preceq T$, and $C \preceq T$.

From L3, $B \preceq T$ and $T \preceq S$ implies that $B \preceq S$; similarly, $C \preceq T$ and $T \preceq S$ implies that $C \preceq S$. Summarizing what we have so far:

$A \preceq S$, $B \preceq S$, and $C \preceq S$.

We will now show that if any other element $S'$ has this same property, then $S \preceq S'$. Assume that

$A \preceq S'$, $B \preceq S'$, and $C \preceq S'$.

From Definition 1 and $B$ cup $C = T$, it follows that $T \preceq S'$; we also know $A \preceq S'$. From Definition 1 and $A$ cup ($B$ cup $C$) $= A$ cup $T = S$, it follows that $A \preceq S$ and $T \preceq S$. Now using the second part of Definition 1, $S \preceq S'$.

In a similar way we now let ($A$ cup $B$) cup $C = U$, and $A$ cup $B = V$; we show that $A \preceq U$, $B \preceq U$, and $C \preceq U$, and that if $U'$ has the same property then $U \preceq U'$. (The details will be asked for in an exercise.)

This means that $S \preceq U$ and $U \preceq S$. Assume that $S \neq U$. Then, from L1, $S \preceq U$ implies that $U \npreceq S$. Since this contradicts $U \preceq S$, our assumption that $S \neq U$ must be false and $S = U$. This means $A$ cup ($B$ cup $C$) = ($A$ cup $B$) cup $C$. ■

## Exercises

1. Show that the definition of $A$ cup $B$ specifies only one object; that is, if $A$ cup $B = C$ and $A$ cup $B = E$, then $C = E$.
2. Complete the proof of the theorem.
3. Prove that $A$ cap ($B$ cap $C$) = ($A$ cap $B$) cap $C$.
4. Prove that $A$ cap $A = A$. How could you prove $A$ cup $A = A$?
5. Prove that $A$ cup $B = B$ cup $A$. How could you prove $A$ cap $B = B$ cap $A$?

*6. In the lattice of Exercise 6 of Section 2, give two examples of $A$ cap ($A$ cup $B$) $= A$. Give two examples of $A$ cup ($A$ cap $B$) $= A$. Prove one of these properties and indicate how the other could be proved.

7. Prove $X$ cup $X = X$ from $A$ cup ($A$ cap $B$) $= A$, and $C$ cap ($C$ cup $D$) $= C$.
8. Let $A \preceq^* B$ mean $A$ cap $B = A$. Assume $A$ cap $A = A$, $A$ cap $B = B$ cap $A$, and ($A$ cap $B$) cap $C = A$ cap ($B$ cap $C$). Prove L1 to L3 for $\preceq^*$.

### 4. AN APPLICATION OF LATTICE THEORY

In the last section we proved that $A$ cup ($B$ cup $C$) = ($A$ cup $B$) cup $C$; that is, the cup operation is associative. Now we can take this valid statement and apply it to any model of lattice theory and it will become a true statement.

In the lattice of divisors of 12 (Section 6, Chapter 1) cup is interpreted as lcm (least common multiple) and lcm ($A$, lcm($B$,$C$)) = lcm (lcm($A$,$B$),$C$) is a true statement.

---

***Show that lcm(4, lcm(3, 6)) = lcm(lcm(4, 3), 6).***

---

To calculate the left-hand side we first find lcm(3,6) = 6, then lcm(4, lcm(3,6)) = lcm(4,6) = 12. In a similar way we can calculate the right-hand side, lcm(4,3) = 12 and lcm(lcm(4,3), 6) = lcm(12,6) = 12.

---

***Find three more examples of the associativity of lcm in the same lattice.***

---

In the lattice of subgroups of Group 4 (Section 6, Chapter 2) cup is interpreted as the "product" of subgroups. We should recall that if **H** and **K** are subgroups, **HK** is the set of all possible products $XY$ in which $X$ is an element of **H** and $Y$ is an element of **K**. The interpretation of the theorem about the associativity of cup is **(HK)L** = **H(KL)** for all normal subgroups of Group 4.

---

***Let* H = {A, J}, K = {A, J, K, L}, *and* L = {A, J, N, Q} *and show that* (HK)L = H(KL).**

---

In the lattice of the seven-point projective geometry (Section 6, Chapter 3) the interpretation of cup is join. If **P, Q, R** are any objects (points, lines, the whole geometry, or the empty object $O$) then (**P** join **Q**) join **R** = **P** join (**Q** join **R**).

---

***Find three examples of the associativity of join in the seven-point projective geometry.***

---

In the lattice of subsets of $\{L, P, R\}$ in this chapter the interpretation of cup is ∪ and the theorem becomes the true statement **A** ∪ (**B** ∪ **C**) = (**A** ∪ **B**) ∪ **C** for all subsets **A, B, C** of $\{L, P, R\}$.

These, of course, are only a few of the models of lattice theory. For every model there will be an interpretation of the theorem which will be a true statement about the associativity of an operation in the model. Our one theorem has given us hundreds (in fact, an infinity) of true statements! This should leave no doubt about the economy of an abstract axiomatic approach to mathematics.

## Exercises

1. The set $\{1,2,3,4,5\}$ under $\leq$ is a lattice (see Exercise 8, Section 2). What is cup? cap? Give three examples to show that the interpretation of cup is associative. Give three examples to show that the interpretation of cap is associative.
2. In the lattice of divisors of 12, what is the interpretation of cap? Give three examples of its associativity.
3. Repeat Exercise 2 for the normal subgroups of Group 4.
4. Repeat Exercise 2 for the lattice of the seven-point projective geometry.
5. Repeat Exercise 2 for the lattice of all subsets of $\{L, P, R\}$.

## 5. WHAT HAVE WE LEARNED FROM LATTICE THEORY?

***Summarize what you have learned from lattice theory.***

The main point of this chapter was to show that it is possible for many seemingly unrelated parts of mathematics to have a *common structure.* Our method of developing axioms for this common structure by looking at several models and trying to characterize them is the way that mathematicians attack the problem of finding an axiom system.

The discovery of this common structure indicates an intrinsic *unity* in mathematics and should give us a feeling of aesthetic satisfaction. The statement, *"Mathematics is beautiful,"* should begin to have some meaning for us after this experience.

The *economy* of the axiomatic approach was also apparent in this chapter. One theorem in abstract lattice theory yields many true statements about the models. The truth of these statements is automatically assured because of their validity in the abstract system. The study of lattice theory thus contributes to many branches of mathematics at the same time. Lattices have been applied in modern algebra, projective geometry, set theory, functional analysis, probability, and logic.

Besides the usefulness of lattice theory within mathematics, it also aids in the solution of many practical problems. One type of lattice which has been very useful is a special one called a Boolean algebra. Boolean algebra has been used in the design of switching circuits. Switching circuits appear in electrical networks, telephone networks, computers, hydraulic systems, and various information systems. A recent application of Boolean algebra is to "neural networks;" Boolean algebra is one of the tools being used to study the large groups of interlinked nerve cells in the brain.

Lattices which are not Boolean algebras also have practical applications. Lattices have been used to study the logic of quantum mechanics, which has applications to wave mechanics. Lattices also aid in the study of fluid turbulence and in the study of the chain reactions that occur in nuclear reactors.

## Exercises

**READ ONE OR BOTH OF THE FOLLOWING**

Lieber, Lillian R., *Lattice Theory,* Galois Institute of Mathematics and Art, Brooklyn, 1959.

The article on Lattice Theory in a recent edition of the *Encyclopaedia Britannica.*

### 6. WHAT IS THE FOUNDATIONS OF MATHEMATICS?

The study of lattices is actually a part of modern algebra. However, since in this chapter we were mainly concerned with developing a set of axioms for lattice theory, we were engaged in an aspect of the foundations of mathematics.

In studying the foundations of mathematics we examine the way in which mathematics is done. Some questions we attempt to answer are:

What is a valid proof?

Which axioms are sufficient for a particular part of mathematics (as lattice theory, or group theory)?

Do two particular different axiom systems represent the same theory (that is, do they have the same models)?

Is a particular axiom system free from contradiction (consistent)?

Can some particular axiom in a system be proved from the rest of the axioms (and so be dropped as an axiom)?

Is there an axiom system sufficient for all of mathematics?

Strictly speaking, these questions are not *within* mathematics but *about* mathematics. The foundations of mathematics rather than being called a branch of mathematics is often called *metamathematics* (beyond or transcending mathematics).

Metamathematicians are concerned about valid proofs, that is, logic. Usually a mathematician takes logic for granted and concentrates on the mathematics he is doing. This is proper, since faster progress is made by isolating parts of a problem. It is one of the jobs of the people in foundations to study the logic used by mathematicians. The next chapter will be about logic.

The study of axiom systems is another part of foundations. Our development of the axioms of lattice theory is typical of this process. The person in foundations may work on the axiom system for one part of mathematics, or he may try to find a system which covers several parts.

At the beginning of this century, it was hoped that an axiom system could be found that would be sufficient for all of mathematics. However, after the work of Kurt Gödel in the 1930's it does not seem that it will be possible to find an axiom system which is consistent (without contradiction) and will be adequate for mathematics in its entirety.

Gödel proved that it is impossible to find a consistent axiom system for even such a simple system as the ordinary arithmetic of the integers. He showed that no matter what consistent system is adopted for arithmetic, it is possible to find a statement about arithmetic which is true but which cannot be proved. Thus Gödel showed that every consistent axiom system for arithmetic is incomplete. If it is not possible to axiomatize arithmetic, it does not seem likely that there is an axiom system for all of mathematics. The hope now is to find a system which will be adequate for a large portion of mathematics and that the parts it does not cover will be of minor importance.

Gödel's work came as a shock to some people. They felt that mathematics had such unity that it could be completely axiomatized. Rather than casting doubt on the unity of mathematics, it seems to indicate that although the axiomatic approach is powerful, it is not rich enough to capture all of the mathematics that the human mind is able to invent. Perhaps the future will bring some other approach. In the meantime, metamathematicians continue to work on axiom systems which cover larger parts of mathematics. One area which is basic to almost every branch of mathematics is set theory, so the study of sets is of much interest to people in foundations. Sets will be the subject of Chapter 6.

## Exercises

**READ ONE OR MORE OF THE FOLLOWING**

Nagel, Ernest, and J. R. Newman, *Gödel's Proof,* New York University Press, New York, 1958. (Also found in *Scientific American,* Vol. 194, No. 6, June, 1956, pp. 71–86; *Mathematics in the Modern World,* pp. 221–230; and *The World of Mathematics,* Vol. 3, pp. 1668–1695.)

*Quine, W. V., "The Foundations of Mathematics," *Scientific American,* Vol. 211, No. 3, September, 1964, pp. 112–127.

Von Mises, Richard, "Mathematical Postulates and Human Understanding," in *The World of Mathematics,* Vol. 3, J. R. Newman, Simon and Schuster, New York, 1956, pp. 1723–1754.

The section about Foundations in the article on Mathematics in a recent edition of the *Encyclopaedia Britannica.*

## Test

**1.** In the set of all subsets of $\{1,2,3\}$, find:

(a) $\{1,2,3\} \cap \{2\}$
(b) $\{1,2,3\} \cap \{\ \}$
(c) $\{1,2,3\} \cap \{1,3\}$
(d) $\{2\} \cap \{2\}$
(e) $\{1,3\} \cup \{\ \}$
(f) $\{2\} \cup \{1,3\}$
(g) $\{1,3\} \cap \{1,3\}$
(h) $\{1,2\} \cap \{2,3\}$
(i) $\{1,2\} \cup \{2,3\}$
(j) $\{\ \} \cup \{\ \}$

**2.** Are the following true or false?

(a) $\{1,2,3\} \subseteq \{1,3\}$
(b) $\{2\} \subseteq \{1,2,3\}$
(c) $(\{2\} \cap \{1,3\}) \subseteq \{\ \}$
(d) $2 \subseteq \{2\}$
(e) $\{\ \} \subseteq \{2\}$

**3.** For the lattice in the sketch are the following true or false?

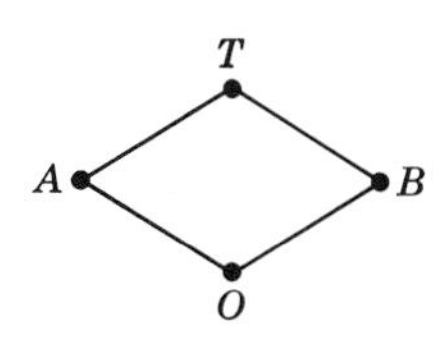


(a) $A \preceq T$
(b) $A \preceq B$
(c) $O \preceq A$
(d) $T \preceq A$
(e) $T \preceq T$

**4.** Using the lattice of Exercise 3 find:

(a) $A$ cup $B$
(b) $A$ cap $B$
(c) $B$ cap $T$
(d) $B$ cap $O$
(e) $A$ cup $T$
(f) $A$ cup $O$
(g) $T$ cup $T$
(h) $O$ cap ($T$ cap $A$)
(i) $A$ cap ($A$ cup $B$)
(j) $T$ cup ($T$ cap $O$)

**5.** Which of the following are lattices?

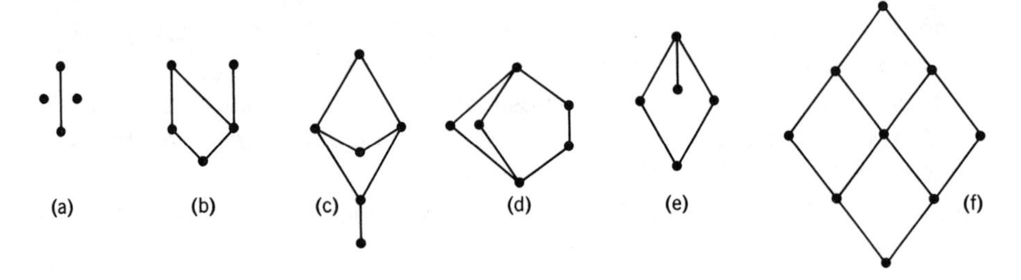


**6.** Find:

(a) gcd(2,8)
(b) gcd(4,6)
(c) gcd(1,10)
(d) gcd(12,18)
(e) gcd(4,4)
(f) lcm(2,3)
(g) lcm(4,8)
(h) lcm(16,24)
(i) lcm(3,5)
(j) lcm(1,10)

**7.** Find the lattice of divisors of 27 ordered by | (divides).

**8.** Explain the relationship between an axiom system and its models.

**9.** Why is it "economical" to study abstract lattice theory?

**10.** What outside reading did you do about the foundations of mathematics? Summarize any facts that you found additional to those in the text.

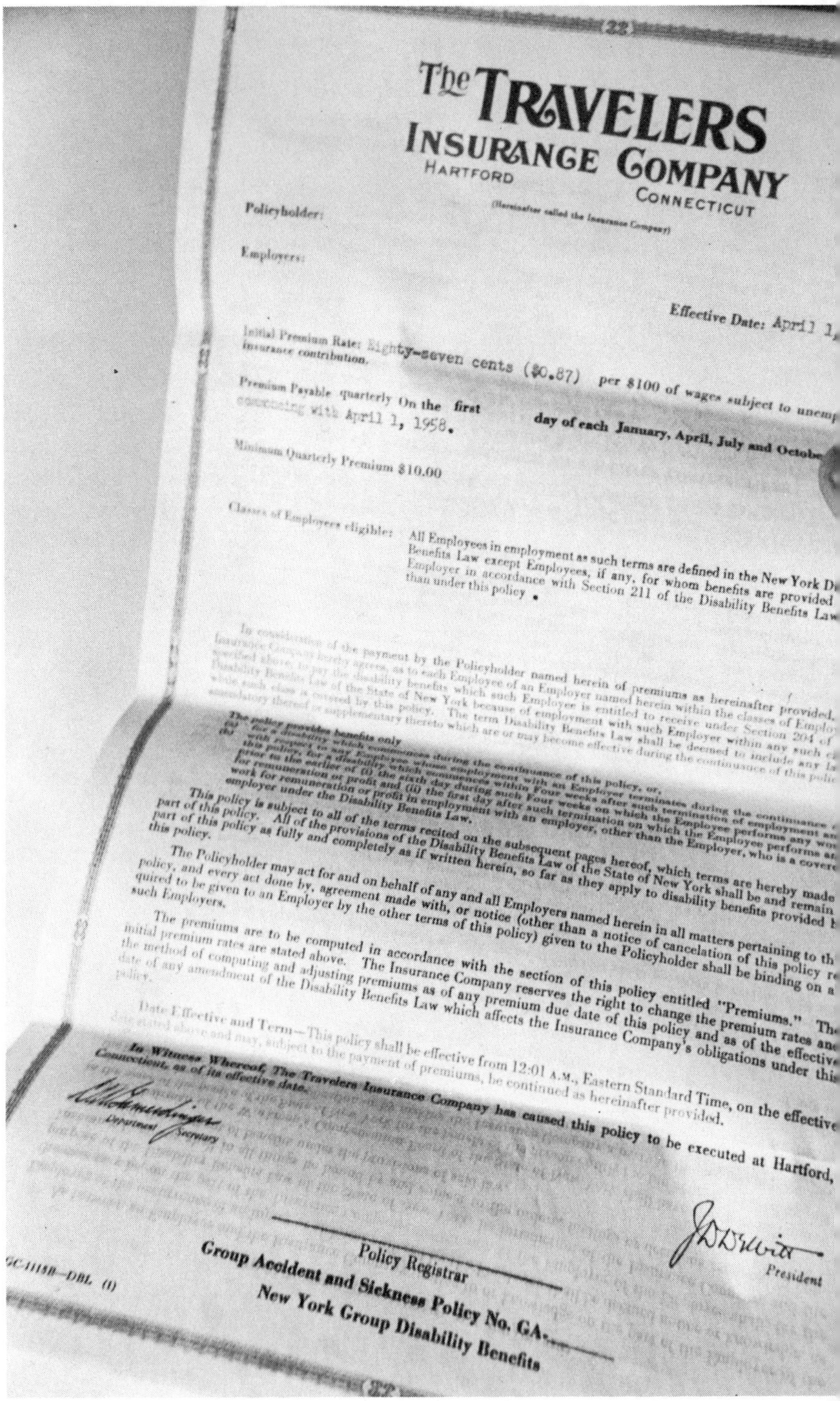

The TRAVELERS
INSURANCE COMPANY
HARTFORD CONNECTICUT
(Hereinafter called the Insurance Company)

Policyholder:

Employers:

Effective Date: April 1,

Initial Premium Rate: Eighty-seven cents ($0.87) per $100 of wages subject to unem
insurance contribution.

Premium Payable quarterly On the first day of each January, April, July and Octobe
commencing with April 1, 1958.

Minimum Quarterly Premium $10.00

Classes of Employees eligible: All Employees in employment as such terms are defined in the New York D
Benefits Law except Employees, if any, for whom benefits are provided
Employer in accordance with Section 211 of the Disability Benefits La
than under this policy.

In consideration of the payment by the Policyholder named herein of premiums as hereinafter provided,
Insurance Company hereby agrees, as to each Employee of an Employer named herein within the classes of Emplo
specified above, to pay the disability benefits which such Employee is entitled to receive under Section 204 of
Disability Benefits Law of the State of New York because of employment with such Employer within any such c
while such class is covered by this policy. The term Disability Benefits Law shall be deemed to include any l
amendatory thereof or supplementary thereto which are or may become effective during the continuance of this polic

The policy provides benefits only
(a) for a disability which commences during the continuance of this policy, or,
(b) for a disability of an Employee whose employment with an Employer terminates during the continuance o
this policy, for a disability which commences within Four weeks after such termination of employment an
prior to the earlier of (i) the sixth day during such Four weeks on which the Employee performs any wo
for remuneration or profit and (ii) the first day after such termination on which the Employee performs a
work for remuneration or profit in employment with an employer, other than the Employer, who is a cover
employer under the Disability Benefits Law.

This policy is subject to all of the terms recited on the subsequent pages hereof, which terms are hereby made
part of this policy. All of the provisions of the Disability Benefits Law of the State of New York shall be and remain
part of this policy as fully and completely as if written herein, so far as they apply to disability benefits provided
this policy.

The Policyholder may act for and on behalf of any and all Employers named herein in all matters pertaining to th
policy, and every act done by, agreement made with, or notice (other than a notice of cancelation of this policy r
quired to be given to an Employer by the other terms of this policy) given to the Policyholder shall be binding on a
such Employers.

The premiums are to be computed in accordance with the section of this policy entitled "Premiums." Th
initial premium rates are stated above. The Insurance Company reserves the right to change the premium rates an
the method of computing and adjusting premiums as of any premium due date of this policy and as of the effectiv
date of any amendment of the Disability Benefits Law which affects the Insurance Company's obligations under thi
policy.

Date Effective and Term—This policy shall be effective from 12:01 A.M., Eastern Standard Time, on the effective
date stated above and may, subject to the payment of premiums, be continued as hereinafter provided.

In Witness Whereof, The Travelers Insurance Company has caused this policy to be executed at Hartford,
Connecticut, as of its effective date.

Secretary

President

Policy Registrar

Group Accident and Sickness Policy No. GA.
New York Group Disability Benefits

GC-1118D—DBL (1)

Van Bucher

5

# MATHEMATICAL LOGIC

## Reasoning about Reasoning

**Logicians are hired to help write insurance policies that are free of inconsistencies and loopholes.**

**W**e have seen that mathematics can take its origin from familiar concrete situations. In Chapter 2 we found groups as an abstraction of rearranging books. We looked at projective geometry in Chapter 3 as an abstraction of casting shadows. We now want to look at another branch of mathematics which begins with a familiar experience. The familiar experience is reasoning, and its abstraction is called mathematical logic.

At first it may not seem possible to study logic, an abstraction of reasoning, in a mathematical way. After all, when we study a branch of mathematics we reason about it. What are we going to use to reason about reasoning? The problem is not insurmountable and we will soon see how we can overcome the difficulty.

## 1. AN EXAMPLE OF REASONING

In order to talk about mathematical logic, the abstraction of reasoning, we must first take a hard look at how we reason. In Sections 1 and 2 we will closely examine some examples of reasoning. In Section 3, we will formulate the mathematical abstraction of reasoning.

Reasoning is a very complex phenomenon. Philosophers, psychologists, semanticists, and logicians all study different aspects of reasoning. The analyses of these scholars have been very useful attempts at understanding reasoning, but we are still far from a full explanation of this very human experience.

We want to look at the one aspect of reasoning which is studied by a logician. The logician is concerned only with the *validity* of reaching a *conclusion* from some accepted statements, called *premises*. An example may help us to understand this.

Assume that a young man is convinced of the truth of the following statements:

PREMISES
1. *I am a pacifist.*
2. *If I am in the military, then I will kill someone.*
3. *I cannot both be a pacifist and kill someone.*

Reasoning from these premises, the young man reaches the following conclusion:

CONCLUSION: *I cannot be in the military.*

The entire process of reaching the conclusion is indeed complex, especially the initial steps of becoming convinced of the truth of the premises. There are many people who would disagree with this man's premises. The logician is of *no help at all* in formulating premises. Is the young man really convinced of the truth of the statements? Are the premises in fact true? These aspects are *not* analyzed by the logician.

The only aspect which the logician looks at is the process of moving from the premises to the conclusion. In this example, the logician would say that the young man's reasoning is *valid;* the conclusion does follow from the premises. (In Section 2 we will analyze why it is valid.) Whether or not the conclusion is *true* is a different question and depends on the truth of the premises. The logician cannot help the man decide on the truth of any of the statements in the argument. In the next section we will see that a person who doubts the truth of the statement must, however, admit that the reasoning is valid.

Let's look at another example of reaching a conclusion from premises:

PREMISES
1. *I am going to be a lawyer.*
2. *If I waste my study time, then I will fail my bar exam.*
3. *I cannot both be a lawyer and fail my bar exam.*

---

***What conclusion seems to follow validly from the above premises?***

---

Although we are probably unable to explain precisely our reasoning, we feel that the conclusion is:

CONCLUSION: *I will not waste my study time.*

---

***Do you see any similarity between these two examples of reasoning?***

---

Although the examples concern completely different situations, the form of the reasoning is identical. This can best be illustrated by using letters *p, q, r* to stand for simple sentences which are in the premises.

| | EXAMPLE 1 | EXAMPLE 2 |
|---|---|---|
| *p* | I am a pacifist. | I am going to be a lawyer. |
| *q* | I am in the military. | I waste my study time. |
| *r* | I will kill someone. | I will fail my bar exam. |

Using these letters (and allowing some leeway for grammar) both of the arguments we have seen have the following form:

PREMISES
1. $p$.
2. If $q$, then $r$.
3. Not both $p$ and $r$.

CONCLUSION: Not $q$.

---

***Give another example of reasoning which has the same form as the last two examples.***

---

There are, of course, many examples. We give only one more:

PREMISES
1. *I want to be in control of my mental processes.*
2. *If a person uses LSD, he loses control of his mental processes.*
3. *It is not possible both to be in control and to lose control of one's mental processes.*

CONCLUSION: *I cannot use LSD.*

## Exercises

**1.**

PREMISES
1. Feathers are soft.
2. If snow is not blue, then roses are green.
3. It is not the case that both feathers are soft and roses are green.

CONCLUSION: Snow is blue.

Does the reasoning above seem valid? Is the conclusion true? Use letters to represent the parts of this argument. How does this pattern of reasoning differ from that in the examples in Section 1?

**2.** Represent the parts of the following argument by letters which stand for simple sentences (you may take some liberties for the sake of grammar).

PREMISES
1. If I take conversational French, I will have to spend time in the language lab.
2. I must take either conversational French or French literature.
3. I am not qualified to take French literature.

CONCLUSION: I will spend time in the language lab.

3. Formulate another argument which has the same structure as the argument in Exercise 2.
4. Represent the parts of the following argument by letters which stand for simple sentences (you may take some liberties for the sake of grammar).

PREMISES {1. I am going to continue dating Ralph.
2. If I marry Fred, then I can no longer date Ralph.

CONCLUSION: I am not going to marry Fred.

5. Formulate another argument which has the same structure as the argument in Exercise 4.

## 2. ANALYZING REASONING

Let's try now to analyze the reasoning used in the examples of Section 1. It is the abstraction of reasoning similar to this which is called logic. In this section we will not yet begin the formal development of logic. The discussion in this section is an attempt to develop our intuitive understanding of reasoning so that the formal development in Section 3 will be meaningful.

In each example of Section 1 we were given some premises. The premises were all declarative sentences. The conclusion which was derived was also a declarative sentence. Logic always deals with *declarative sentences* because they make a direct statement which imparts information. It would be much more difficult to reach conclusions from questions (as "Is it raining?") or from commands (as "Open the door!"), although often even in these cases some information is imparted.

Sometimes the declarative sentences in reasoning are *simple* (rather than compound) *positive* (no "not," "no," or their equivalent in the sentence) *sentences.* Simple positive sentences are found in 1 of each example: "I am a pacifist." "I am going to be a lawyer." "I want to be in control of my mental processes." Other times the sentences are compound and we can think of them as made up of simple sentences as in 2 and 3. In 2 the simple positive sentences $q$ and $r$ are connected by *if* ______, *then* ______. For example, "I am in the military," and "I will kill someone," are connected to form, "*If* I am in the military, *then* I will kill someone." (Sometimes "if $q$, then $r$" is written as "*$q$ implies $r$.*") In 3 the simple positive sentences $p$ and $r$ are connected by *and* and then the entire expression is negated by *not.* For example, "I am a pacifist," and "I will kill someone" are connected to form "I am a pacifist *and* I will kill someone," and then this is negated to form the sentence "I can*not* both be a pacifist and kill someone." (Once again we must remind ourselves not to get overconcerned about minor changes in the sentences that are necessary for the sake of grammar.) In other examples of reasoning, *or* is commonly used to connect sentences. The words "and," "or," and "implies" (or "if ______, then ______") are called *connectives.* Although "not" is not used to put sentences together, but rather to negate sentences, it is also called a connective.

Each declarative sentence we reason about is either *true* or *false.* We do not necessarily know which of these cases holds, but we do assume that at least one of them does in fact hold. We also assume that the sentence is not both true and false at the same time. These two conditions mean that we exclude from our reasoning statements like the isolated sentence "It is five feet tall" and "This sentence is false."

***Why do you think these last two sentences are excluded from reasoning?***

"It is five feet tall," would be a perfectly acceptable sentence with which to reason if we knew the antecedent of "it." We cannot reason with the isolated sentence because alone it is neither true nor false.

"This sentence is false," is a paradoxical sentence. If it is true, then it says of itself that it is false. If it is false, then it says that it is true. We feel that sentences of this type which are both true and false are impossible to reason with, so we do not allow their use in an argument. We reason only with sentences which are either true or false but not both.

Let's summarize now what we have noticed about our reasoning process. Reasoning is concerned with declarative sentences. These sentences can be positive simple sentences or they may be made up of positive simple sentences by using the connectives "and," "or," "implies," and "not." The only sentences which are considered are those which are either true or false, but not both true and false.

Besides these rather elementary notions about reasoning, the important thing is that somehow we move from a given set of premises to the conclusion. Let's look at the first example once again.

PREMISES
1. *I am a pacifist.*
2. *If I am in the military, then I will kill someone.*
3. *I cannot both be a pacifist and kill someone.*

CONCLUSION: *I cannot be in the military.*

***Try to explain why you feel the conclusion is valid.***

We might say that from 1 and 3, we conclude that "I will not kill anyone." "I will not kill anyone," combined with 2 leads us to the conclusion "I cannot be in the military."

We reason so often and so quickly that it is difficult to explain why we reason as we do. The explanation in the paragraph above is far from complete. How do we conclude from 1 and 3 that "I will not kill anyone"? And then how do we move from "I will not kill anyone" and 2 to the conclusion?

---

***Try to explain how 1 and 3 lead to "I will not kill anyone."***

---

One explanation would be the following. We might first say that 3, "I cannot both be a pacifist and kill someone," implies "If I am a pacifist, then I will not kill anyone." We feel that some principle or *law of reasoning* is involved in this implication and we are willing to accept "If I am a pacifist, then I will not kill anyone" because we have accepted 3. Then since we have "I am a pacifist" as premise 1, we accept the sentence "I will not kill anyone."

Two times in the above explanation we had an if ______, then ______ expression, accepted the *if* clause, and broke off (or *detached)* the *then* clause. The processes of detaching the *then* clause when the *if* clause is a premise or has been reasoned to, seems to be some sort of *rule* of logic.

---

***Try to explain how "I will not kill anyone" and 2 lead to the conclusion.***

---

This time we might say that 2, "If I am in the military, then I will kill someone," implies "If I will not kill anyone, then I cannot be in the military," by some law of logic. Then since we have accepted the *if* clause, "I will not kill anyone," we may detach the *then* clause, "I cannot be in the military."

The explanation we have just given is difficult to follow. It is much easier to analyze the reasoning if we replace the parts of the premises by letters as in Section 1.

| | |
|---|---|
| $p$ | I am a pacifist. |
| $q$ | I am in the military. |
| $r$ | I will kill someone. |

The entire reasoning process could then be summarized as follows:

| | |
|---|---|
| A. Not both $p$ and $r$ *implies* if $p$ then not-$r$. | (LAW OF REASONING) |
| B. Not both $p$ and $r$. | (PREMISE 3) |
| C. If $p$, then not-$r$. | (A, B, AND DETACHMENT) |
| D. $p$. | (PREMISE 1) |
| E. Not-$r$. | (C, D, AND DETACHMENT) |

| | |
|---|---|
| F. If $q$ then $r$ *implies* if not-$r$ then not-$q$. | (LAW OF REASONING) |
| G. If $q$, then $r$. | (PREMISE 2) |
| H. If not-$r$, then not-$q$. | (F, G, AND DETACHMENT) |
| I. Not-$q$. | (H, E, AND DETACHMENT) |

## Exercises

**1.** Solve the following puzzle. (In puzzles we often assume silly statements, but this lets us concentrate on the validity of the argument instead of the truth of the premises.)

Bob likes blondes.
Tom does not take out Sue implies that Mary is a redhead.
It is not the case that both Bob likes blondes and Mary is a redhead.
Pair up Bob and Tom with Mary and Sue.

**2.** Analyze your reasoning in the solution of Exercise 1.

**3.** Solve the following puzzle. Match three boys Ken, Bill, and Jim with the cars Ford, Olds, and Chevy if:

A. It is not the case that both Jim is not a senior and Ken does not own a Chevy.
B. Jim is a senior implies that Ken owns a Chevy.
C. Either Ken is a sophomore or he does not own a Chevy.
D. Bill does not drive a Ford implies that Ken is not a sophomore.

***4.** Solve the following puzzle. Mr. and Mrs. Smith and Mr. and Mrs. Jones go out to dinner together. Each has a different preference as to how he likes his steak to be served. One prefers rare, another medium-rare, another medium, and another well-done. Of the four, two are from the East and one of these likes steak medium-rare. There is only one wife from the East; she likes her steak well-done and her husband likes his rare. Mr. Jones is from the Midwest. What is each person's preference in steaks?

### 3. AXIOMS FOR A TWO-VALUED LOGIC OF SENTENCES

Historically, the task of formulating an abstraction of reasoning was not a smooth process. In fact, it took centuries before it was achieved. The ancient Greeks at the time of Aristotle began to collect some of the principles or laws of reasoning and to study how they are related to each other. It was not until the 19th century that a logic for sentences was finally fully organized. Since this logic is an abstraction of reasoning in which a sentence is either true or false, it is called a *two-valued logic of sentences.*

Let's take a brief look at this modern abstraction of reasoning. A two-valued logic of sentences deals with some basic objects, $p, q, r, \ldots$ which are called *atoms* (the abstraction of simple positive sentences). The atoms are joined together with $\wedge$ read "and," $\vee$ read "or" (and meaning "and/or"), $\rightarrow$ read "implies," and $\neg$ read "not" to form WFFs, *well-formed formulas* (the abstraction of simple and compound sentences). There are three rules which tell us how to recognize WFFs:

R1. Any atom is a WFF.
R2. If $A$ is a WFF, then $\neg A$ is a WFF.
R3. If $A$ and $B$ are WFFs, then $A \wedge B$, $A \vee B$, and $A \rightarrow B$ are WFFs.

The rules for forming WFFs are an abstraction of the part of grammar which tells us how to form correct compound sentences.

The laws of reasoning that we saw in the examples of the first two sections and similar laws will be the axioms (WFFs we assume) and theorems (WFFs we prove from the axioms) of mathematical logic. In our formal development we shall call both the axioms and theorems *laws.*

The following set of axioms are often used for a two-valued logic of sentences:

For all WFFs $A, B, C,$ the following are laws:

A1. $A \rightarrow (B \rightarrow A)$
A2. $[A \rightarrow (B \rightarrow C)] \rightarrow [(A \rightarrow B) \rightarrow (A \rightarrow C)]$
A3. $A \rightarrow [B \rightarrow (A \wedge B)]$
A4. $(A \wedge B) \rightarrow A$
A5. $(A \wedge B) \rightarrow B$
A6. $A \rightarrow (A \vee B)$
A7. $B \rightarrow (A \vee B)$
A8. $(A \rightarrow C) \rightarrow \{(B \rightarrow C) \rightarrow [(A \vee B) \rightarrow C]\}$
A9. $(A \rightarrow B) \rightarrow [(A \rightarrow \neg B) \rightarrow \neg A]$
A10. $\neg \neg A \rightarrow A$

We list these formidable looking axioms to illustrate one approach to two-valued logic. We are not going to explain all of the axioms or even show that they are reasonable abstractions of everyday reasoning. We certainly do not suggest memorizing them. The only point of this list is that logicians begin developing logic by accepting certain WFFs as laws and this list is an example of the WFFs they might assume.

Some of these axioms are easy to understand and obviously reflect everyday reasoning. If applied to reasoning, A4 means "If both $A$ and $B$ are true, then $A$ alone is true." Law A5 has a similar meaning. Applied to reasoning, A6 means "If $A$ is true, then either $A$ or $B$ or possibly both are true." A similar thing is meant by A7. Axiom A10 means that double negation implies an affirmation, a common principle in grammar. A summary of what happens in an argument by contradiction is contained in A9.

If A1 is applied to reasoning it means "If $A$ is true, then *anything* (that is, $B$) implies $A$." This seems like a rather silly statement to make. We must agree, however, that at least it will not lead us to anything false, since $A$ is true anyway. This first axiom is very useful when proving theorems as we will soon see. It must be admitted, however, that its presence in this axiom system makes it less than a perfect abstraction of reasoning. (There are other systems of logic which have been developed in an attempt to overcome this difficulty.)

We will accept A2, A3, and A8 as axioms of this system even though it is not immediately apparent why they are abstractions of everyday reasoning. Although we use everyday reasoning for inspiration, abstract mathematical logic does not depend on the correspondence of our axioms to the reasoning process. We may assume anything we want in an abstract system as long as it does not lead to a contradiction; then we concentrate on the validly derived theorems.

Theorems are WFFs which we derive validly from the axioms. This is the point at which we must decide how we are going to reason about mathematical logic. Certainly we do not want to reason in our usual way; this would be circular and would not give us any insight into everyday reasoning. Instead of reasoning as usual, we will restrict our derivations to one rule called the *rule of detachment.*

D. *If $A \rightarrow B$ is a law and $A$ is a law, then $B$ is a law.*

Theorems are derived as follows. A proof of a theorem will consist of a list of WFFs. The last WFF in the list is the WFF we want to show is a law. Each of the WFFs in the list (including the last one) is either an axiom, a previously proved theorem, or a WFF which has been obtained from two previous WFFs in the list by means of D, the rule of detachment. An example will help us understand this.

T1. $X \rightarrow X$ (Applied to reasoning, this means "Any sentence implies itself.")

*Proof*

1. $X \rightarrow [(X \vee Y) \rightarrow X]$ (This is A1 with $X = A$, $X \vee Y = B$. Since $X \vee Y$ is a WFF it can be used in place of $B$).
2. $\{X \rightarrow [(X \vee Y) \rightarrow X]\} \rightarrow \{[X \rightarrow (X \vee Y)] \rightarrow (X \rightarrow X)\}$ (This is A2 with $X = A$, $X \vee Y = B$, $X = C$.)

3. $[X \rightarrow (X \vee Y)] \rightarrow (X \rightarrow X)$ (This is derived from 1 and 2 by D.)
4. $X \rightarrow (X \vee Y)$ (This is A6 with $X = A$ and $Y = B$.)
5. $X \rightarrow X$ (This is derived from 3 and 4 by D.) ■

It is not too difficult to see that this proof does satisfy the description given above. However, we might wonder how anyone ever finds a proof of a theorem. It often requires much ingenuity and patience. In practice, the logician often works backward to construct the proof. Let's look at another proof.

T2. $\neg (X \wedge \neg X)$ (Applied to reasoning this means "It is not the case that a sentence and its negation are true at the same time.")

*Proof*

| | | |
|---|---|---|
| 1. | $[(X \wedge \neg X) \rightarrow X] \rightarrow \{[(X \wedge \neg X) \rightarrow \neg X] \rightarrow \neg(X \wedge \neg X)\}$ | (A9) |
| 2. | $(X \wedge \neg X) \rightarrow X$ | (A4) |
| 3. | $[(X \wedge \neg X) \rightarrow \neg X] \rightarrow \neg (X \wedge \neg X)$ | (D, 1, 2) |
| 4. | $(X \wedge \neg X) \rightarrow \neg X$ | (A5) |
| 5. | $\neg (X \wedge \neg X)$ | (D, 3, 4) ■ |

***Explain which WFFs were used in place of* A *and* B *in the axioms.***

In the proof of T2, $(X \wedge \neg X) = A$ and $X = B$ in A9, $X = A$ and $\neg X = B$ in A4, and $X = A$ and $\neg X = B$ in A5.

We'll give one final example. This time, supplying the reasons will be left as an exercise for the reader.

T3. $(X \wedge Y) \rightarrow (X \vee Y)$ (This applied to reasoning means "If two sentences are true then it is correct to say that at least one of them is true.")

*Proof*

1. $[X \rightarrow (X \vee Y)] \rightarrow \{(X \wedge Y) \rightarrow [X \rightarrow (X \vee Y)]\}$
2. $X \rightarrow (X \vee Y)$
3. $(X \wedge Y) \rightarrow [X \rightarrow ( X \vee Y]$
4. $\{(X \wedge Y) \rightarrow [X \rightarrow (X \vee Y)]\} \rightarrow \{[(X \wedge Y) \rightarrow X] \rightarrow [(X \wedge Y) \rightarrow (X \vee Y)]\}$
5. $[(X \wedge Y) \rightarrow X] \rightarrow [(X \wedge Y) \rightarrow (X \vee Y)]$
6. $(X \wedge Y) \rightarrow X$
7. $(X \wedge Y) \rightarrow (X \vee Y)$ ■

Besides these theorems there are many other laws of this logic. Many of them are very difficult to prove. A few simple ones will be suggested in the exercises. It would be nice if there were an easier way to determine if a given WFF was a law of logic. There is such a method and it will be discussed in the next section.

## Exercises

**1.** Which of the following are WFFs?

(a) $\neg AB$

(b) $A \rightarrow$

(c) $\neg\neg\neg A$

(d) $\neg(\neg A \vee B)$

(e) $(A \rightarrow B) \rightarrow \neg A$

(f) $[(A \vee B) \vee C] \vee D$

(g) $(A \rightarrow B) \vee (\rightarrow C)$

(h) $(A \rightarrow B) \vee \neg C$

(i) $A \neg B$

(j) $\vee A \rightarrow B$

**2.** Make up a sentence which has the structure of A3.

**3.** Give all the reasons in the proof of T3.

***4.** Prove $(X \rightarrow \neg X) \rightarrow \neg X$ using T1 and A9.

***5.** Prove $(Y \wedge X) \rightarrow (X \vee Y)$ using a proof similar to T3.

### 4. TWO-VALUED TRUTH TABLES

We have just seen in the last section that a WFF is a law of reasoning if it can be derived from a certain set of axioms. In practice this is a very difficult way to have to decide whether or not a given WFF is a law. If it is a law it might take us a very long time to discover a proof from the axioms. If some given WFF is not a law, the situation is even worse; we would not be able to find a proof, but we wouldn't know whether we hadn't worked long enough or if there simply was no proof.

Fortunately there is a much easier way to decide which WFFs are laws. The technique which is used is called a *truth table.* We assume that every WFF has a truth value of either true or false. We shall use 1 to stand for true and 0 for false.

In ordinary reasoning we feel that if a sentence is true, then its negation is false; and if a sentence is false, then its negation is true. This corresponds to the following table in our abstract system:

| $A$ | $\neg A$ |
|---|---|
| 1 | 0 |
| 0 | 1 |

We also feel that an "and" sentence is true if both parts of it are true, and is false in all other cases. This corresponds in the abstract system to the following table:

| $\wedge$ | 1 | 0 |
|---|---|---|
| 1 | 1 | 0 |
| 0 | 0 | 0 |

The left-hand column of the table represents the truth value of $A$ in the WFF $A \wedge B$, the top row represents the truth value of $B$, and the truth values for the entire WFF $A \wedge B$ appear in the body of the table.

The table for $\vee$ can be written in a similar way. We feel that an "or" sentence, since "or" means "and/or", is false only if both of its parts are false. In the abstract system this corresponds to:

| $\vee$ | 1 | 0 |
|---|---|---|
| 1 | 1 | 1 |
| 0 | 1 | 0 |

The truth table for $\rightarrow$ is a little harder to accept as an abstraction of everyday reasoning.

| $\rightarrow$ | 1 | 0 |
|---|---|---|
| 1 | 1 | 0 |
| 0 | 1 | 1 |

Applied to reasoning it means that the only time a sentence with "implies" in it is false is if the sentence is made up of a true statement implying a false statement. We would certainly agree that a sentence of this type is false. It is harder to understand why something false implying something true is true, and why something false implying something false is true. The best we can say is that we certainly do not feel that sentences of this type are false. For example, if I say

"If I get paid today, I'll give you a dollar,"

and in fact I do not get paid today, then whether or not I do give you a dollar you cannot accuse me of lying. My sentence was not false, so we say that it is true. (This is certainly another weakness of two-valued logic. It would be better to say that the truth value of my sentence cannot be determined, or perhaps that it is *maybe*. In the next section we will look at another logic in which this is possible.)

Using these four tables we can now find the truth table for any WFF; for example, $\neg A \vee A$.

| ¬ | A | ∨ | A |
|---|---|---|---|
| 0 | 1 | 1 | 1 |
| 1 | 0 | 1 | 0 |
| ③—① | | | ② |
| | | ④ | |

The steps we should follow are:

①Under the first letter which appears from the left list both truth values.

②Under any other occurrence of this letter list the truth values again in the same order.

③Since ¬ $A$ must be evaluated before we can get the truth value of the entire expression, we fill in this column with the negations of column ①.

④Finally, consulting the truth table for ∨, we fill in this column by applying ∨ to columns ③ and ②.

Column ④ lists the truth values for the entire expression ¬ $A \lor A$. In this particular case the WFF is always true. A WFF which is true (has value 1) for all substitutions of 1 and 0 is called a *tautology.*

***Find the truth table for ¬ A → A.***

We should find this table as follows:

| ¬ | A | → | A |
|---|---|---|---|
| 0 | 1 | 1 | 1 |
| 1 | 0 | 0 | 0 |
| ③—① | | | ② |
| | | ④ | |

The WFF ¬ $A \to A$ is *not* a tautology because the WFF is not always true.

If a WFF has two different variables (letters) in it, then we will need four rows to compute all possible combinations of truth values for the two variables. For example, the truth table for (¬ $A \lor B) \to (A \to B)$ is:

| (¬ | A | ∨ | B) | → | (A | → | B) |
|---|---|---|---|---|---|---|---|
| 0 | 1 | 1 | 1 | 1 | 1 | 1 | 1 |
| 0 | 1 | 0 | 0 | 1 | 1 | 0 | 0 |
| 1 | 0 | 1 | 1 | 1 | 0 | 1 | 1 |
| 1 | 0 | 1 | 0 | 1 | 0 | 1 | 0 |
| ⑤—① | | | ③ | | ② | | ④ |
| | | ⑥ | | | | ⑦ | |
| | | | | ⑧ | | | |

① Under the first letter which appears from the left, list the truth values 1, 1, 0, 0.

② Under any other occurrence of this letter list the same truth values.

③ Under the second letter which appears from the left, list the truth values 1, 0, 1, 0.

④ Under any other occurrence of this second letter repeat the same truth values as in step ③.

⑤ Apply ¬ to the truth values of column ①.

⑥ Apply ∨ to columns ⑤ and ③.

⑦ Apply → to columns ② and ④.

⑧ Apply → to ⑥ and ⑦.

$(\neg A \vee B) \rightarrow (A \rightarrow B)$ is a tautology.

***Find the truth table for*** $(\neg A \vee B) \rightarrow \neg B$.

| (¬ | A | ∨ | B) | → | ¬ | B |
|---|---|---|---|---|---|---|
| 0 | 1 | 1 | 1 | 0 | 0 | 1 |
| 0 | 1 | 0 | 0 | 1 | 1 | 0 |
| 1 | 0 | 1 | 1 | 0 | 0 | 1 |
| 1 | 0 | 1 | 0 | 1 | 1 | 0 |
| ④—① | | ⑥ | ② | ⑦ | ⑤—③ | |

WFFs with three letters will require eight rows. The steps in the following example should be clear from the circled numbers.

| [A | → | (B | → | C)] | → | [(A | → | B) | → | (A | → | C)] |
|---|---|---|---|---|---|---|---|---|---|---|---|---|
| 1 | 1 | 1 | 1 | 1 | 1 | 1 | 1 | 1 | 1 | 1 | 1 | 1 |
| 1 | 0 | 1 | 0 | 0 | 1 | 1 | 1 | 1 | 0 | 1 | 0 | 0 |
| 1 | 1 | 0 | 1 | 1 | 1 | 1 | 0 | 0 | 1 | 1 | 1 | 1 |
| 1 | 1 | 0 | 1 | 0 | 1 | 1 | 0 | 0 | 1 | 1 | 0 | 0 |
| 0 | 1 | 1 | 1 | 1 | 1 | 0 | 1 | 1 | 1 | 0 | 1 | 1 |
| 0 | 1 | 1 | 0 | 0 | 1 | 0 | 1 | 1 | 1 | 0 | 1 | 0 |
| 0 | 1 | 0 | 1 | 1 | 1 | 0 | 1 | 0 | 1 | 0 | 1 | 1 |
| 0 | 1 | 0 | 1 | 0 | 1 | 0 | 1 | 0 | 1 | 0 | 1 | 0 |
| ① | ⑨ | ④ | ⑧ | ⑥ | ⑬ | ② | ⑩ | ⑤ | ⑫ | ③ | ⑪ | ⑦ |

The interesting thing about truth tables is that every law of the last section (every axiom and every theorem) is a tautology, and every tautology is a law. This means that, instead of proving laws from the axioms, we can quickly determine whether or not a given WFF is a law by looking at its truth table to see whether or not it is a tautology.

Truth tables have another use. If the truth values of two WFFs are the same for corresponding substitutions of 1 and 0, as in the following tables (notice the same substitutions are made for *A* and the same for *B*)

| $A$ | $\wedge$ | $\neg$ | $B$ |
|---|---|---|---|
| 1 | 0 | 0 | 1 |
| 1 | 1 | 1 | 0 |
| 0 | 0 | 0 | 1 |
| 0 | 0 | 1 | 0 |
| ① | ④ | ③ | ② |

| $\neg$ | $(\neg$ | $A$ | $\vee$ | $B)$ |
|---|---|---|---|---|
| 0 | 0 | 1 | 1 | 1 |
| 1 | 0 | 1 | 0 | 0 |
| 0 | 1 | 0 | 1 | 1 |
| 0 | 1 | 0 | 1 | 0 |
| ⑤ | ③ | ① | ④ | ② |

we say that the two WFFs are *equivalent.* Applied to reasoning, if the truth values of the two WFFs which correspond to two sentences are the same, we can substitute one sentence for the other.

The equivalence above tells us that the sentence

> "I am taking English 1313 and I am not taking History 1613,"

and the sentence

> "It is not the case that: I am not taking English 1313 or I am taking History 1613,"

are equivalent.

***Show that A $\rightarrow$ B and $\neg$ A $\vee$ B are equivalent. Make up two sentences which illustrate these WFFs.***

## Exercises

**1.** Show that Axioms A1 to A10 of the last section are all tautologies.

**2.** Show that Theorems T1 to T3 of the last section are all tautologies.

**3.** Which of the following are tautologies?

(a) $\neg A \wedge B$
(b) $(A \vee B) \vee C$
(c) $(\neg A \vee B) \rightarrow \neg (A \wedge \neg B)$
(d) $A \vee \neg A$
(e) $\neg (\neg A \vee B)$
(f) $(A \wedge B) \rightarrow (B \wedge A)$
(g) $A \rightarrow \neg (A \rightarrow \neg A)$
(h) $(A \rightarrow \neg A) \rightarrow A$
(i) $A \wedge \neg A$
(j) $(A \rightarrow B) \rightarrow (\neg A \rightarrow \neg B)$
(k) $(\neg A \rightarrow A) \rightarrow A$
(l) $A \rightarrow \neg \neg A$
(m) $(A \vee B) \rightarrow (B \vee A)$

**4.** Show that the following WFFs are equivalent:

$\neg \neg A$ and $A$,
$A \vee B$ and $(A \rightarrow B) \rightarrow B$,
$\neg \neg \neg A$ and $\neg A$,
$A \rightarrow B$ and $\neg B \rightarrow \neg A$,
$A \wedge B$ and $\neg (\neg A \vee \neg B)$,
$\neg A \vee B$ and $\neg (A \wedge \neg B)$.

## 5. THREE-VALUED TRUTH TABLES

We might wonder why we developed a two-valued logic, a logic in which well-formed formulas are abstractions of sentences which are either true or false. Real life situations are usually not so black and white. Perhaps it would be more realistic to assign three truth values to WFFs, true, false, and maybe. For example, the sentence "I hate rhubarb" could be other than completely true, or completely false. If in fact I merely tolerate rhubarb, then it is not true that I hate rhubarb, nor is it false that I hate rhubarb. Something in between seems more appropriate.

In 1920, Lukasiewicz suggested a three-valued logic. In this logic WFFs are formed as before and the usual connectives are used. WFFs, however, may have three truth values: true, maybe, and false. Let true be represented by 1, maybe by $1/2$, and false by 0.

As in the two-valued development we can set up truth tables. In Exercise 4 of the last section we found that $A \vee B$ is equivalent to $(A \rightarrow B) \rightarrow B$ and that $A \wedge B$ is equivalent to $\neg (\neg A \vee \neg B)$. Because of this it seems that if we could find three-valued truth tables for $\neg$ and $\rightarrow$, we could use these to figure out the tables for $\vee$ and $\wedge$.

We will find the three-valued truth tables by generalizing the two-valued tables. The two-valued table for $\neg$ was

| $A$ | $\neg A$ |
|---|---|
| 1 | 0 |
| 0 | 1 |

We could have expressed this as a mathematical formula, $\neg A = 1 - A$.

***Check that this formula is equivalent to the table for $\neg$.***

Using this formula, the three-valued truth table for $\neg$ is

| $A$ | $\neg A$ |
|---|---|
| 1 | 0 |
| $1/2$ | $1/2$ |
| 0 | 1 |

In a similar way, if we could find a mathematical formula for the two-valued $\rightarrow$ table, we could use it to find the more general three-valued table for $\rightarrow$.

***Try to find a formula for two-valued $\rightarrow$.***

A formula is not as easy to find as the one for $\neg$. Lukasiewicz looked at the two-valued table for $\rightarrow$,

| $\rightarrow$ | 1 | 0 |
|---|---|---|
| 1 | 1 | 0 |
| 0 | 1 | 1 |

and noticed that $A \rightarrow B$ is 1 if $A \leq B$, and $A \rightarrow B$ is $1 - A + B$ if $A > B$. Using this formula, the three-valued table for $\rightarrow$ is

| $\rightarrow$ | 1 | $1/2$ | 0 |
|---|---|---|---|
| 1 | 1 | $1/2$ | 0 |
| $1/2$ | 1 | 1 | $1/2$ |
| 0 | 1 | 1 | 1 |

***Using the three-valued tables for $\neg$ and $\rightarrow$ and the fact that* A $\vee$ B *is equivalent to* (A $\rightarrow$ B) $\rightarrow$ B, *find the three-valued truth table for $\vee$.***

The truth table for $\vee$ is:

| $\vee$ | 1 | $1/2$ | 0 |
|---|---|---|---|
| 1 | 1 | 1 | 1 |
| $1/2$ | 1 | $1/2$ | $1/2$ |
| 0 | 1 | $1/2$ | 0 |

If we try to express the three-valued table for $\vee$ in terms of a mathematical formula we see that $A \vee B$ is always the greater of $A$ and $B$. If we look at the two-valued table for $\vee$ in the last section, we see that this is also so. This is encouraging, for we seem to have a generalization of two-valued logic which preserves some of its basic properties.

---

***Using the three-valued truth tables for $\neg$ and $\vee$ and the fact that $\mathbf{A} \wedge \mathbf{B}$ is equivalent to $\neg(\neg \mathbf{A} \vee \neg \mathbf{B})$, find the three-valued truth table for $\wedge$.***

---

We should find the following table:

| $\wedge$ | 1 | ½ | 0 |
|---|---|---|---|
| 1 | 1 | ½ | 0 |
| ½ | ½ | ½ | 0 |
| 0 | 0 | 0 | 0 |

The mathematical formula this time is that $A \wedge B$ is the smaller of $A$ and $B$. Again this property is shared by the two-valued $\wedge$ table.

We could now ask which WFFs are three-valued tautologies.

---

***Show that $\mathbf{A} \rightarrow \mathbf{A}$, $(\mathbf{A} \wedge \mathbf{B}) \rightarrow (\mathbf{B} \wedge \mathbf{A})$, and $\mathbf{A} \rightarrow \neg\neg \mathbf{A}$ are three-valued tautologies.***

---

We might notice that the three-valued tautologies just mentioned are also two-valued tautologies. In fact, every three-valued tautology is a two-valued tautology since the two-valued table is a part of the three-valued table.

---

***Show that $\mathbf{A} \vee \neg \mathbf{A}$, $\mathbf{A} \rightarrow \neg(\mathbf{A} \rightarrow \neg \mathbf{A})$ and $(\neg \mathbf{A} \rightarrow \mathbf{A}) \rightarrow \mathbf{A}$ are not three-valued tautologies.***

---

All three of these WFFs are two-valued tautologies, however, they are not tautologies in three-valued logic. We should not begin to think, however, of three-valued logic as just a piece of two-valued logic. In a three-valued system we can make distinctions which we cannot make in a two-valued system. For example, let's define $\square A$ to be $\neg A \rightarrow A$. We can read $\square A$ as, "$A$ is possible."

***Find both the two- and the three-valued tables for □ A.***

| $A$ | $\Box A$ |
|---|---|
| 1 | 1 |
| 0 | 0 |

| $A$ | $\Box A$ |
|---|---|
| 1 | 1 |
| ½ | 1 |
| 0 | 0 |

In the two-valued table for $\Box A$, $A$ and $\Box A$ have the same truth values and so are equivalent. In the three-valued table for $\Box A$, this is not the case; $A$ and $\Box A$ are not equivalent. In a two-valued logic it is not possible to distinguish $A$ from $\Box A$, but in a three-valued logic this distinction can be made.

It is possible to generalize even further and consider $n$-valued logics. For example, if we want a 6-valued logic we could use truth values 1, 4/5, 3/5, 2/5, 1/5, and 0, and use the same formulas for the three tables:

$$\neg A = 1 - A$$

$$A \rightarrow B = \begin{cases} 1 & \text{if } A \leq B \\ 1 - A + B & \text{if } A > B \end{cases}$$

$$A \vee B = \text{the greater of } A \text{ and } B$$

$$A \wedge B = \text{the smaller of } A \text{ and } B$$

Many-valued logics originated as attempts to formulate a logic that would reflect everyday reasoning more accurately than does the two-valued system. At the present time, they are not used to test the validity of everyday reasoning. They have been used, however, as a tool for the study of two-valued logics. Many-valued logics are also used in the computer solutions of some special problems. It has been suggested that an interpretation of quantum mechanics in a many-valued logic would avoid some of the difficulties that arise when describing the results of quantum mechanics in a two-valued logic. Although the applications at present are not extensive, many-valued logics are interesting in themselves as generalizations of two-valued logic. Besides this reason for studying many-valued logics, developments in this area may be preparing an important mathematical tool for future scientists.

## Exercises

**1.** Which of the axioms A1 to A10 for two-valued logic are three-valued tautologies?

**2.** Find the truth tables for $\neg$, $\rightarrow$, $\vee$, and $\wedge$ in a four-valued logic.

***3.** Find three more two-valued tautologies which are not three-valued tautologies.

## 6. WHAT HAVE WE LEARNED FROM THE LOGICS OF SENTENCES?

Our study of mathematical logic points out once again that mathematics usually begins with some *concrete* situation. We started with everyday reasoning, then formulated an abstract system of logic similar to it.

No one system (two-valued logic) will be a perfect picture of the real world, so *different abstractions* (such as three-valued logic) are developed in an attempt to better reflect some aspect of the real world.

Once we have developed an abstraction it is not necessary to keep comparing it with the real world. Theorems will be valid in our system if we can prove them from our basic assumptions and not on the basis of some conformity with the real world. We can even *generalize* our abstraction until it seems to resemble the real situation very little as in the case of $n$-valued logic (or non-Euclidean geometries in Chapter 3).

Finally, our study of logic teaches us that mathematics deals with things other than numbers, or things resembling numbers, and things other than geometrical figures. Mathematical logic was not an abstraction of numbers or space, but an abstraction of reasoning.

Recently several new branches and applications of mathematics have developed which do not abstract from numbers or geometrical figures. Some of these are game theory, automata, information theory, cybernetics, and linguistics. These fields could be called applied mathematics because they are used to try to solve some problems in the real world. In the process, most of them have had to develop a considerable amount of new mathematics to fit the situation with which they are dealing.

In *game theory* we try to analyze the problems of conflict. We are concerned with situations which involve strategy rather than just chance, so are interested in games like poker and chess rather than bingo. In game theory we abstract the common features of strategy. Game theory was begun in 1920 and, although it applies to games, it was developed as a means of dealing with competitive economic behavior. Not only economics, but also sociology, psychology, politics, and war have gamelike aspects for which game theory may be helpful.

*Automata* is the mathematical abstraction of automated systems. For example, it deals with features which are common to the combination lock and the computer. Although automata does not deal directly with the design of computers, it proves general results which are helpful in their design.

*Information theory* is the mathematical abstraction of communications.

Information theory considers the symbols used in communication, but is so general that it does not say what these symbols are. They could be written or spoken letters or words, musical notes or symphonic music, or even pictures. The relationships information theory discovers apply to all these and other forms of communications.

*Cybernetics,* which emerged in 1948, is the comparative study of automated systems. Cybernetics looks for common elements in the functioning of automated machines and the human nervous system. It tries to develop a theory which will cover the entire field of control and communications in machines and living organisms.

*Mathematical linguistics* isolates the structural properties that are common to all languages and describes them mathematically. Deductive reasoning is then used to derive consequences from these basic properties. In mathematical linguistics we are trying to understand the nature of human language and perhaps contribute to an understanding of human thought. Mathematical linguistics has been useful in programming computers to do translating.

## Exercises

**READ ONE OR MORE OF THE FOLLOWING**

Harris, Zellig, "Mathematical Linguistics," *The Mathematical Sciences,* edited by COSRIMS, M.I.T. Press, Cambridge, Mass., 1969, pp. 190–196.

*Morgenstern, Oskar, "The Theory of Games," *Scientific American,* Vol. 180, No. 5, May, 1949, pp. 22–25.

*Rapoport, Anatol, "The Use and Misuse of Game Theory," *Scientific American,* Vol. 207, No. 6, December, 1962, pp. 108–118.

*Weaver, Warren, "The Mathematics of Communication," *Scientific American,* Vol. 181, No. 1, July, 1949, pp. 11–15.

*Wiener, Norbert, "Cybernetics," *Scientific American,* Vol. 179, No. 5, November, 1948, pp. 14–19.

An *Encyclopaedia Britannica* (recent edition) article on: Cybernetics; Games, the Theory of; or Information Theory.

### 7. WHAT IS LOGIC?

The study of logic began with Aristotle. Men, of course, reasoned long before Aristotle, but he was the first to study the principles of reasoning.

As in our development, he restricted his discussion to declarative sentences and divided these into simple and composite sentences. *Unlike* our development of logic, he looked at the inner structure of the sentences and spoke of predicating things of a subject. He distinguished sentences which had a particular subject, as "Socrates is mortal," from a universal subject "Every man is mortal," and from the less general subject "Some men are lazy." He further distinguished positive and negative sentences. He then listed valid ways of deducing things from known things (syllogisms) and even used letters (variables) to stand for the subjects and predicates in his sentences.

Aristotle would have accepted the following as a valid way of reasoning:

> If $A$ is true of all $B$, and $B$ is true of all $C$, it is necessary for $A$ to be true of all $C$.

A particular example of this would be:

> All animals are mortal and all men are animals, so it is necessary that all men are mortal.

Aristotle accepts certain syllogisms as axioms and proves that other syllogisms are valid from these, or proves that they are not valid by counterexamples. Aristotle's work is the first use of variables and the first use of axioms (it preceded the axiomatic development of geometry).

About the time of Euclid a more sophisticated logic appeared, the logic of the Stoics. Their variables stood for entire sentences as ours do and it was very much like our modern mathematical logic. They listed various valid arguments and described how to derive others from them. One of their valid arguments was:

> The first or the second;
> Not the first;
> Therefore, the second.

For centuries after the Stoics there were few logicians worthy of mention. However, a fragment of the ancient logic was preserved and passed on. About the 12th century, after centuries of being virtually lost, the full Aristotelian logic became known to the Scholastics. They relearned it, developed another logic of their own, and synthesized the two. Scholastic logic depended heavily on Latin, the language of the schools, and reached its peak during the 14th and 15th centuries. As Latin declined, this logic too was lost, and not found again until the middle of this century.

In the meantime, mathematical logic as it is known today began its growth in the last part of the 19th century. Besides the logics of sentences which we looked at, it also includes tools to handle the internal parts of sentences. It covers more complex types of reasoning than any of the previous logics, and in a certain sense scholastic logic and Aristotelian logic are part of modern mathematical logic.

Mathematical logic has many uses besides the obvious one of being a tool to analyze reasoning. It has been used by insurance companies to develop a rule for arranging premium payments by policyholders (see Pfeiffer's article on "Symbolic Logic" referenced in the exercises), and to analyze their policies. Large corporations also use logic to analyze and simplify their contracts and to check them for loopholes and inconsistencies. Logic can also be used to check the accuracy of censuses and polls.

Besides these applications several others arise, because two-valued logic of sentences is another model of a lattice which is a Boolean algebra (see Section 5 of Chapter 4). All of the applications of Boolean algebra can be thought of as applications of the two-valued logic of sentences.

## Exercises

**READ ONE OR MORE OF THE FOLLOWING**

Lewis, Clarence I., and Cooper H. Langford, "History of Symbolic Logic," in *The World of Mathematics,* Vol. 3, J. R. Newman, Simon and Schuster, New York, 1956, pp. 1859–1877.

Lieber, Lillian R., *Mits, Wits, and Logic,* Third Edition, Norton, New York, 1960.

Nagel, Ernest, "Symbolic Notation, Haddocks' Eyes and the Dog-Walking Ordinance," in *The World of Mathematics,* Vol. 3, J. R. Newman, Simon and Schuster, New York, 1956, pp. 1878–1900.

*Pfeiffer, J. E., "Symbolic Logic," *Scientific American,* Vol. 183, No. 6, December, 1950, pp. 22–24.

Tarski, Alfred, "Symbolic Logic," in *The World of Mathematics,* Vol. 3, J. R. Newman, Simon and Schuster, New York, 1956, pp. 1901–1931.

The article on Logic in a recent edition of the *Encyclopaedia Britannica.*

## Test

1. Which of the following are WFFs?

(a) $A \neg$
(b) $(A \rightarrow B) \rightarrow$
(c) $\neg B \rightarrow A$
(d) $\rightarrow A \wedge B$
(e) $[(A \wedge B) \wedge A] \vee B$
(f) $\neg \neg C \neg$
(g) $\neg \neg \neg \neg \neg C$
(h) $\wedge \wedge C$
(i) $\neg (\neg A \wedge \neg B \wedge)$
(j) $(A \rightarrow B) \rightarrow (B \rightarrow A)$
(k) $\wedge A \rightarrow B$
(l) $[(A \wedge B) \vee (C \wedge B)] \rightarrow \neg A$

2. Three of our axioms for two-valued logic were:

I. $A \rightarrow (B \rightarrow A)$
II. $(A \wedge B) \rightarrow A$
III. $A \rightarrow [B \rightarrow (A \wedge B)]$

For each of the following identify which axiom above it is and which WFFs have been used in place of $A$ and $B$.

(a) $(X \wedge Y) \rightarrow \{W \rightarrow [(X \wedge Y) \wedge W]\}$
(b) $(X \wedge Y) \rightarrow [(X \rightarrow Y) \rightarrow (X \wedge Y)]$
(c) $[(X \vee Y) \wedge (Y \rightarrow Z)] \rightarrow (X \vee Y)$

3. Find the two-valued truth tables for:

(a) $\neg (A \wedge \neg A)$
(b) $\neg A \vee \neg B$
(c) $A \rightarrow \neg B$
(d) $\neg A \wedge \neg B$
(e) $(A \rightarrow B) \rightarrow A$
(f) $\neg (A \wedge B)$

4. Which of the WFFs in Question 3 are tautologies?

5. List any equivalent WFFs in Question 3.

6. Write out a five-valued truth table for $\rightarrow$.

7. What did we learn about mathematics from our study of logic?

8. What outside reading did you do about logic, game theory, automata, information theory, or cybernetics? Summarize any facts you found which were additional to the text.

R. Gates-Frederic Lewis, Inc.

6

# SET THEORY

## Finding an Infinity of Infinite Numbers

The abstract approach of set theory led to a new branch of mathematics, functional analysis, which has been used to arrive at the streamlined design of this jet.

**M**ost people are familiar with the notions of set and subset. It would be incorrect, however, to say that they know about set theory. Although the notion of set is a very simple intuitive concept, some of the theorems which can be proved about sets are among the least intuitive facts of mathematics. In this chapter we will look at an unintuitive part of set theory—infinite cardinal numbers.

## 1. COMPARING THE CARDINALITIES OF SETS

Let's imagine that we are small children. We can speak in short sentences, but cannot count yet. In front of us are

three dolls,
one balloon,
five crayons,
six blocks, and
five toy cars.

We are asked, "Are there more balloons than dolls?" We quickly answer, "No, more dolls." Somehow we have already developed the notions of *one,* and *more than one.* Then we are asked, "Are there more cars than blocks?" We do not know how to answer. We can't count, and the number of cars looks almost the same as the number of blocks.

***How could a child compare the number of cars with the number of blocks without counting?***

One possibility would be to line them up side by side so that each car is next to one block. The double arrow indicates that two objects correspond. There would be a block which is not next to a car, so there are more blocks than cars.

If we had been asked to compare the number of crayons with the number of cars, we could proceed in a similar way. This time each car would correspond to one crayon, and each crayon to one car. We would answer that the number of cars is the same as the number of crayons. It was not necessary to count or know about numbers.

These examples are almost ridiculously simple, yet they contain some basic concepts of set theory, which lead to very interesting results. We'll

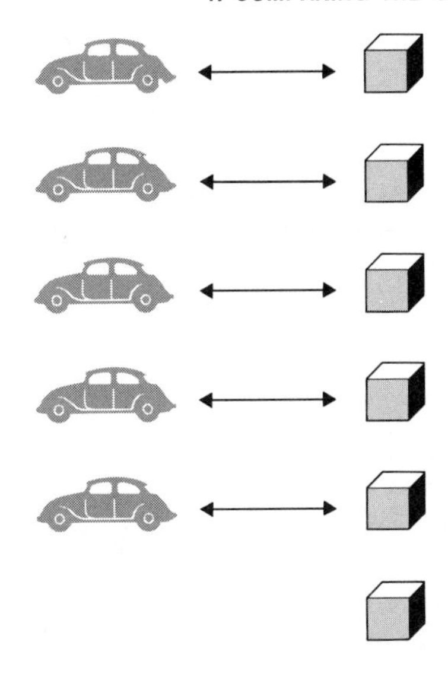

clarify the basic concepts now; the rest of the chapter will be concerned with the facts that follow from these concepts.

If **A** and **B** are any two sets, we might want to know how their sizes compare, that is, how the number of elements in **A** compare with the number of elements in **B**. We would like to make this comparison, however, without using numbers. Mathematicians call the size of a set the *cardinality* of a set. Let's write # (**X**) for the cardinality of the set **X**.

***Formulate a definition for #(A) = #(B) without using numbers. (Hint: Describe mathematically the process of the child.)***

An acceptable definition would be:

> DEFINITION: Two sets **A** and **B** have the same cardinality, #(**A**) = #(**B**), if and only if, there is a correspondence such that every element of **A** corresponds to exactly one element of **B**, and every element of **B** corresponds to exactly one element of **A**. (This is called a *1-to-1 correspondence.*)

It might be helpful to look at some examples of equal cardinality.

***If C = {a,b,c}, D = {8,5,2}, and E = {x,y}, show that #(C) = #(D) and #(C) ≠ #(E).***

The cardinalities of **C** and **D** are the same because a 1-to-1 correspondence is possible, for example:

$$a \leftrightarrow 5$$
$$b \leftrightarrow 2$$
$$c \leftrightarrow 8$$

The cardinalities of **C** and **E** are *not* the same because every correspondence fails to be 1-to-1. For example, in this correspondence

$$a \leftrightarrow y$$
$$b \leftrightarrow x$$
$$c \swarrow\!\!\nearrow$$

every element of **C** is associated with exactly one element of **E**, however, $x$ in **E** corresponds to two elements of **C**. A similar thing happens no matter what correspondence between these two sets we consider.

The child was able to do better than say that two sets were of unequal size, he also could figure out which one was larger (or which was smaller). Mathematically we are able to do the same thing. If the cardinality of **A** is less than the cardinality of **B**, we'll write #(**A**) < #(**B**). What we need now is a definition of this notion.

---

***Formulate a definition for #(A) < #(B) without using numbers. (Hint: Use the process of the small child.)***

---

Remembering the child's method of comparing the number of cars with the number of blocks we could formulate the following definition.

> DEFINITION: The cardinality of set **A** is less than the cardinality of set **B**, #(**A**) < #(**B**), if and only, if the cardinality of **A** is equal to the cardinality of a subset of **B** and the cardinality of **B** is not equal to the cardinality of any subset of **A**.

If **C** $= \{a,b,c\}$ and **E** $= \{x,y\}$, then we can write #(**E**) < #(**C**) because #(**E**) $= \#(\{a,c\})$, $\{a,c\} \subseteq$ **C**, but #(**C**) is not equal to the cardinality of any subset of **E**.

For small sets, the notion of equal cardinality and < for cardinality do not tell us anything new. We could have reached the same conclusion by counting, and since counting is more familiar to us it would have been easier. Our definitions seem to be belaboring the obvious. If we consider large sets, however, the comparisons are much more difficult to make and the need for the definitions becomes clear.

We are going to consider twelve sets of various sizes. We will sometimes use the notation **Y** = {**X** | **X** is . . . .} in the following discussion. This notation is read, "**Y** is the set of all **X** such that **X** is . . . ." Notice in particular that " | " means "such that" in this context and *not* "divides" as in Chapter 1.

The twelve sets we are going to consider are the following:

| Set | Description in Words |
|---|---|
| $\mathbf{A} = \{1,2,3,\ldots\}$ | The set of all natural numbers. |
| $\mathbf{B} = \{P,Q,R,S\}$ | The set of the four upper case letters $P,Q,R,$ and $S$ of the English alphabet. |
| $\mathbf{C} = \{a,b,c,\ldots,y,z\}$ | The set of all the lower case letters in the English alphabet. |
| $\mathbf{D} = \{\mathbf{X} \mid \mathbf{X} \subseteq \mathbf{A}\}$ | The set of all **X** such that **X** is a subset of **A**, that is, the set of all subsets of the set of natural numbers. |
| $\mathbf{E} = \{10, 100, 1000, \ldots\}$ | The set of all positive powers of 10. |
| $\mathbf{F} = \{\mathbf{X} \mid \mathbf{X} \subseteq \mathbf{C}\}$ | The set of all **X** such that **X** is a subset of **C**, that is, the set of all subsets of the set of lower case letters in the English alphabet. |
| $\mathbf{G} = \{\mathbf{X} \mid \mathbf{X} \subseteq \mathbf{B}\}$ | The set of all subsets of the set which contains $P,Q,R,$ and $S$. |
| $\mathbf{H} = \{n \mid n = 0.a_1a_2a_3\ldots$ and each $a$ is one of $0,1,2,\ldots,9\}$ | The set of all real numbers between and including 0 and 1. |
| $\mathbf{I} = \{2,4,6,\ldots\}$ | The set of all positive even numbers. |
| $\mathbf{J} = \{a/b \mid a$ and $b$ are natural numbers$\}$ | The set of all positive fractions. |
| $\mathbf{K} = \{\mathbf{X} \mid \mathbf{X}$ is a person alive in 1960$\}$ | The set of all persons alive in 1960. |
| $\mathbf{L} = \{\mathbf{X} \mid \mathbf{X} \subseteq \mathbf{H}\}$ | The set of all subsets of the set of real numbers between and including 0 and 1. |

***Name four members of each of the preceding twelve sets. Make a statement about the number of elements in each set.***

Our examples of members and statements about the number of elements could be as follows:

| Set | Some Members | Size |
|---|---|---|
| **A** | 4, 10, 101, 23647 | infinite (that is, a list of the elements would not end) |
| **B** | *P, Q, R, S* | 4 members |
| **C** | *f, h, k, t* | 26 members |
| **D** | {1,3}, {2,3,4,5}, { }, {3,6,9,...} | infinite |
| **E** | $10^4$, $10^7$, $10^{10}$, $10^{215}$ | infinite |
| **F** | {*c*}, {*a,d,s*}, {*z,t*}, {*k,l,m,n,o*} | more than 26 members |
| **G** | { }, {*P*}, {*S*}, {*P,R*} | 16 members |
| **H** | 0.001, 0.999 . . . , 0.1313 . . . , 0.225 | infinite |
| **I** | 10, 18, 216, 1332 | infinite |
| **J** | 1/3, 5/9, 121/7, 3/2 | infinite |
| **K** | John Kennedy, Jane Fonda, John Glenn, Lew Alcindor | very large |
| **L** | { }, {0.6, 0.35}, {0.1, 0.11, 0.111, . . .}, {0.111 . . . , 0.222 . . . , 0.333 . . . , . . . , 0.999 . . .} | infinite |

***Use the definitions of equal cardinality and < for cardinality to try to arrange these twelve sets according to size beginning with the smallest.***

The task of arranging these sets according to size is not a simple one. We can begin the list now, but it will not be completed in this or even the next section.

***Find the first four sets in the arranged list beginning with the smallest.***

Since for small sets our definitions give us the same results as counting, it is not hard to pick out the first four sets. The smallest set in the list is certainly $\mathbf{B} = \{P, Q, R, S\}$. The next is $\mathbf{G} = \{\mathbf{X} \mid \mathbf{X} \subseteq \mathbf{B}\}$, since **G** contains the sixteen subsets of **B** (see Section 1, Chapter 4). The set of letters in the English alphabet, $\mathbf{C} = \{a, b, c, \ldots, y, z\}$, is third on the list. The set of all subsets of **C**, $\mathbf{F} = \{\mathbf{X} \mid \mathbf{X} \subseteq \mathbf{C}\}$, is next. (The number of elements in **F** is $2^{26}$. The proof of this is left as an exercise to be done at the end of this section.)

Our list begins

$$\begin{aligned} \mathbf{B} &= \{P, Q, R, S\} \\ \mathbf{G} &= \{\mathbf{X} \mid \mathbf{X} \subseteq \mathbf{B}\} \\ \mathbf{C} &= \{a, b, c, \ldots, y, z\} \\ \mathbf{F} &= \{\mathbf{X} \mid \mathbf{X} \subseteq \mathbf{C}\} \end{aligned}$$

because the number of elements in these sets are 4, 16, 26, $2^{26}$, respectively. Even if we didn't know how to count the elements in each of these sets, we could still arrange them according to size because $\#(\mathbf{B}) < (\mathbf{G}) < \#(\mathbf{C}) < \#(\mathbf{F})$ from the definition of $<$ for cardinality of sets. For example $\#(\mathbf{B}) < \#(\mathbf{G})$ because $\#(\mathbf{B}) = \#(\mathbf{M})$ where $\mathbf{M} = \{\{P\}, \{P, Q\}, \{\ \}, \{Q, R, S\}\}$, $\mathbf{M} \subseteq \mathbf{G}$, and $\mathbf{G} = \{\mathbf{X} \mid \mathbf{X} \subseteq \mathbf{B}\}$ does not have the same cardinality as any subset of **B**. The actual carrying out of all the details of this approach, however, would be quite tedious.

***Find the fifth set in the list.***

There is one set which differs from all the rest which remain to be arranged. The set **K**, a very large set, is finite while all the rest are infinite. We don't know exactly how large **K** is (an educated guess is available), but there is certainly some natural number $m$ such that $\#(\mathbf{K}) = \#(\{1, 2, 3, \ldots, m\})$. We say that **K** is *finite* because its cardinality is equal to the cardinality of a set of the form $\{1, 2, 3, \ldots, m\}$ where $m$ is a natural number. A set is *infinite* if it is not finite. Intuitively, a set is infinite if it is always possible to name another element no matter how many elements have already been named. The rest of the sets, **A, D, E, H, I, J, L** are all infinite.

Our list now looks as follows:

$$\begin{aligned} \mathbf{B} &= \{P, Q, R, S\} \\ \mathbf{G} &= \{\mathbf{X} \mid \mathbf{X} \subseteq \mathbf{B}\} \end{aligned}$$

$$\mathbf{C} = \{a,b,c,\ldots,y,z\}$$
$$\mathbf{F} = \{\mathbf{X} \mid \mathbf{X} \subseteq \mathbf{C}\}$$
$$\mathbf{K} = \{\mathbf{X} \mid \mathbf{X} \text{ is a person alive in 1960}\}$$

and then some arrangement of the infinite sets **A, D, E, H, I, J,** and **L**.

The theory of infinite sets was first developed at the end of the 19th century by Georg Cantor. Since no one has actually seen an infinite set in its entirety, the study of such sets is some distance from the real physical world. In spite of this lack of tangible evidence for the existence of infinite sets, they are acceptable objects of study for most mathematicians and are used unhesitatingly in most branches of mathematics.

## Exercises

1. List all of the subsets of:
   (a) $\{\ \}$ (b) $\{1\}$ (c) $\{1,2\}$ (d) $\{1,2,3\}$ (e) $\{1,2,3,4\}$ (f) $\{1,2,3,4,5\}$
2. Use the answers to Exercise 1 to make a conjecture about the number of subsets of $\{1,2,3,\ldots,n\}$.
3. Prove that the number of subsets of $\mathbf{C} = \{a,b,c,\ldots,y,z\}$ is $2^{26}$.
4. Show that if $\mathbf{G} = \{\mathbf{X} \mid \mathbf{X} \subseteq \mathbf{B}\}$, $\mathbf{B} = \{P,Q,R,S\}$, and $\mathbf{C} = \{a,b,c,\ldots,y,z\}$, then there is a set **M**, $\mathbf{M} \subseteq \mathbf{C}$, and $\#(\mathbf{G}) = (\mathbf{M})$.
5. If $\mathbf{A} = \{a\}$ and $\mathbf{B} = \{c,d\}$ show that the definition for $\#(\mathbf{A}) < \#(\mathbf{B})$ is satisfied.

## 2. THE CARDINALITY OF THE SET OF NATURAL NUMBERS

In the last section we saw that sets **A, D, E, H, I, J,** and **L** were all infinite sets. We still must decide how to arrange them according to cardinality if indeed they can be.

***Arrange A, E, I, J in order according to cardinality beginning with the smallest.***

All four of these sets are infinite,

$$\mathbf{A} = \{1,2,3,\ldots\}$$
$$\mathbf{E} = \{10,\ 100,\ 1000,\ldots\}$$
$$\mathbf{I} = \{2,4,6,\ldots\}$$
$$\mathbf{J} = \{a/b \mid a \text{ and } b \text{ are natural numbers}\}.$$

Some of these sets are subsets of others, in fact $\mathbf{E} \subseteq \mathbf{I} \subseteq \mathbf{A}$ because every positive power of 10 is a positive even number and every positive even number is a natural number. Since every natural number $x$ can be written as the fraction $x/1$, then **A** has the same cardinality as a subset of **J**. Our intuition tells us that $\#(\mathbf{E}) < \#(\mathbf{I}) < \#(\mathbf{A}) < \#(\mathbf{J})$—but our intuition is WRONG! Let's consider sets **I** and **A**. If we look at the correspondence

| | | | | | | | | | |
|---|---|---|---|---|---|---|---|---|---|
| **A**: | 1 | 2 | 3 | 4 | 5 | 6 | 7 | 8 | ... |
| | | $\updownarrow$ | | $\updownarrow$ | | $\updownarrow$ | | $\updownarrow$ | |
| **I**: | | 2 | | 4 | | 6 | | 8 | ... |

we do have $\#(\mathbf{I}) = \#(\mathbf{M})$ where **M** is a subset of **A**. However, it is also possible to consider a correspondence $n \leftrightarrow 2^n$

| | | | | | | | | | | |
|---|---|---|---|---|---|---|---|---|---|---|
| **A**: | 1 | 2 | | 3 | | | | 4 | | ... |
| | $\updownarrow$ | $\updownarrow$ | | $\updownarrow$ | | | | $\updownarrow$ | | |
| **I**: | 2 | 4 | 6 | 8 | 10 | 12 | 14 | 16 | 18 | ... |

in which $\#(\mathbf{A}) = \#(\mathbf{N})$ where **N** is a subset of **I**. This last correspondence shows that the definition of $<$ for cardinality is *not* satisfied for **I** and **A**.

***How do the cardinalities of A and I compare?***

The correct answer is that $\#(\mathbf{A}) = \#(\mathbf{I})$. This can be seen by the 1-to-1 correspondence $n \leftrightarrow 2n$

| | | | | | | |
|---|---|---|---|---|---|---|
| **A**: | 1 | 2 | 3 | 4 | 5 | ... |
| | $\updownarrow$ | $\updownarrow$ | $\updownarrow$ | $\updownarrow$ | $\updownarrow$ | |
| **I**: | 2 | 4 | 6 | 8 | 10 | ... |

***How do the cardinalities of E and A compare?***

Again we have an unintuitive fact, $\#(\mathbf{E}) = \#(\mathbf{A})$ because of the 1-to-1 correspondence $n \leftrightarrow 10^n$

| | | | | | | |
|---|---|---|---|---|---|---|
| **A**: | 1 | 2 | 3 | 4 | 5 | ... |
| | $\updownarrow$ | $\updownarrow$ | $\updownarrow$ | $\updownarrow$ | $\updownarrow$ | |
| **E**: | 10 | 100 | 1000 | 10000 | 100000 | ... |

The fact that the set of natural numbers, the set of even natural numbers, and the set of positive powers of ten are all the same size is nothing short of amazing. The even numbers **I** are a *proper subset* of the natural numbers **A**, that is, $\mathbf{I} \subseteq \mathbf{A}$ and $\mathbf{I} \neq \mathbf{A}$. Similarly, the positive powers of ten **E**, are a *proper subset* of the natural numbers, $\mathbf{E} \subseteq \mathbf{A}$ and $\mathbf{E} \neq \mathbf{A}$.

The 1-to-1 correspondence between **A** and **I** means that we could throw away half of all the natural numbers (the odd numbers) and still have as many numbers as we started with. Further, the 1-to-1 correspondence of **A** and **E** means that we could throw away all of the natural numbers except the "few" which are powers of 10 and still have as many as we started with! By considering the negation of the concept of finite sets (infinite sets) we have moved from one of the most primitive concepts man has (comparing the sizes of sets) to a fact which defies all intuition—for infinite sets, a part (proper subset) of a set need not be smaller than the whole set.

***Compare the cardinalities of A and J.***

We have seen enough amazing things to conjecture that $\#(\mathbf{A}) = \#(\mathbf{J})$. To prove this we must find 1-to-1 correspondence between **A** and **J**. This can be accomplished as follows. The elements of **J** are of the form $a/b$ with $a$ and $b$ natural numbers. Let $a/b \leftrightarrow 2^a \cdot 3^b$; for example

$$1/2 \leftrightarrow 2^1 \cdot 3^2 = 18$$
$$1/3 \leftrightarrow 2^1 \cdot 3^3 = 54$$
$$2/3 \leftrightarrow 2^2 \cdot 3^3 = 108$$
$$2/5 \leftrightarrow 2^2 \cdot 3^5 = 972$$
$$7/3 \leftrightarrow 2^7 \cdot 3^3 = 3456$$
$$5/4 \leftrightarrow 2^5 \cdot 3^4 = 2592$$

The correspondence above is 1-to-1 between **J** and an infinite subset **T** of the natural numbers. (Show that **T** is a proper subset of the natural numbers.) A second 1-to-1 correspondence, between **T** and the natural numbers **A**, can be obtained by listing the elements of **T** in order according to size beginning with the smallest and then associating the first number with 1, the second with 2, and so on. Combining these two 1-to-1 correspondences (between **J** and **T**, and between **T** and **A**) we get a 1-to-1 correspondence between sets **J** and **A** which shows that $\#(\mathbf{J}) = \#(\mathbf{A})$.

Let's write out our list of sets now according to cardinality as we can see it at the present time.

$$\mathbf{B} = \{P, Q, R, S\}$$
$$\mathbf{G} = \{\mathbf{X} \mid \mathbf{X} \subseteq \mathbf{B}\}$$
$$\mathbf{C} = \{a, b, c, \ldots, y, z\}$$

$\mathbf{F} = \{\mathbf{X} \mid \mathbf{X} \subseteq \mathbf{C}\}$
$\mathbf{K} = \{\mathbf{X} \mid \mathbf{X} \text{ is a person alive in 1960}\}$

$$\begin{cases} \mathbf{A} = \{1, 2, 3, \ldots\} \\ \mathbf{E} = \{10, 100, 1000, \ldots\} \\ \mathbf{I} = \{2, 4, 6, \ldots\} \\ \mathbf{J} = \{a/b \mid a \text{ and } b \text{ are natural numbers}\} \end{cases}$$

We group **A, E, I,** and **J** together to indicate that they have equal cardinality. We still do not know where to place

$\mathbf{D} = \{\mathbf{X} \mid \mathbf{X} \subseteq \mathbf{A}\}$
$\mathbf{H} = \{n \mid n = 0.a_1a_2a_3 \ldots \text{ and each } a \text{ is one of } 0, 1, \ldots, 9\}$
$\mathbf{L} = \{\mathbf{X} \mid \mathbf{X} \subseteq \mathbf{H}\}$.

These last three sets are certainly infinite, but are they the same size as **A, E, I,** and **J**? We will continue the problem in the next section.

## Exercises

Show that all of the following sets have the same cardinality by showing that each set has the same cardinality as $\mathbf{A} = \{1, 2, 3, \ldots\}$.

1. $\mathbf{M} = \{3, 4, 5, \ldots\}$
2. $\mathbf{N} = \{0, 1, 2, 3, \ldots\}$
3. $\mathbf{O} = \{\ldots, -2, -1, 0, 1, 2, \ldots\}$
4. $\mathbf{P} = \{3, 6, 9, 12, \ldots\}$
5. $\mathbf{Q} = \{-1, -2, -3, \ldots\}$
6. $\mathbf{R} = \{1, 4, 9, 16, \ldots\}$
7. $\mathbf{S} = \{1, 8, 27, \ldots\}$

*8. $\mathbf{T} = \{a/b \mid a \text{ and } b \text{ are integers and } b \neq 0\}$

### 3. THE CARDINALITY OF THE POWER SET OF A SET

In the previous section we considered the sets $\mathbf{B} = \{P, Q, R, S\}$ and $\mathbf{G} = \{\mathbf{X} \mid \mathbf{X} \subseteq \mathbf{B}\}$. The set **G** of all subsets of the set **B** is called the *power set* of **B**. We will write $\mathscr{P}(\mathbf{B})$ for the power set of **B**.

We have already made some comparisons between the cardinality of a set and the cardinality of its power set. We know that $\#(\mathbf{B}) < \#(\mathscr{P}(\mathbf{B}))$ because **B** contains four elements and $\mathscr{P}(\mathbf{B})$ contains sixteen. For $\mathbf{C} = \{a, b, c, \ldots, y, z\}$ we have $\#(\mathbf{C}) < \#(\mathscr{P}(\mathbf{C}))$ since $\mathscr{P}(\mathbf{C})$ contains $2^{26}$ elements. In the case of finite sets it is not hard to show that $\#(\mathbf{X}) < \#(\mathscr{P}(\mathbf{X}))$. If **X** contains $n$ elements then $\mathscr{P}(\mathbf{X})$ contains $2^n$ elements.

What about infinite sets? The set $\mathbf{A} = \{1, 2, 3, \ldots\}$ and $\mathbf{D} = \{\mathbf{X} \mid \mathbf{X} \subseteq \mathbf{A}\} = \mathscr{P}(\mathbf{A})$ are among the sets to be arranged according to cardinality.

***How does the cardinality of A compare with the cardinality of*** $\mathbf{D} = \mathscr{P}(\boldsymbol{A})$***?***

At this point we should be a little afraid to generalize what happens in the finite case. We have seen that facts which are obvious about finite sets need not be true for infinite sets; besides, our first impression is that all infinite sets are the same size. However, our experience with finite sets is the only thing we have to inspire conjectures. It is not unreasonable to conjecture that for *all* sets $\mathbf{X}$, $\#(\mathbf{X}) < \#(\mathscr{P}(\mathbf{X}))$, but of course we must hold this conjecture under suspicion until we prove that it is true.

The conjecture is in fact true and was first proved by Cantor. The proof we will now develop is similar to his.

THEOREM OF CANTOR: For all sets $\mathbf{X}$, $\#(\mathbf{X}) < \#(\mathscr{P}(\mathbf{X}))$.

*Proof:* We must use the definition of $<$ for cardinality of sets and show that $\mathbf{X}$ has the same cardinality as some subset of $\mathscr{P}(\mathbf{X})$, and that $\mathscr{P}(\mathbf{X})$ does not have the same cardinality as any subset of $\mathbf{X}$.

There is one subset of $\mathscr{P}(\mathbf{X})$ which is a natural one to set in 1-to-1 correspondence with $\mathbf{X}$, the subset $\mathbf{Y} = \{\{x\} \mid x \text{ is in } \mathbf{X}\}$. The correspondence $x \leftrightarrow \{x\}$ is 1-to-1 between $\mathbf{X}$ and $\mathbf{Y}$ so $\#(\mathbf{X}) = \#(\mathbf{Y})$. (For example, if $\mathbf{X} = \{a,b,c\}$, $\mathscr{P}(\mathbf{X}) = \{\{\ \}, \{a\}, \{b\}, \{c\}, \{a,b\}, \{a,c\}, \{b,c\}, \{a,b,c\}$, $\mathbf{Y} = \{\{a\}, \{b\}, \{c\}\}$, $\mathbf{Y} \subseteq \mathscr{P}(\mathbf{X})$ and $a \leftrightarrow \{a\}$, $b \leftrightarrow \{b\}$, $c \leftrightarrow \{c\}$.)

In order to show that $\mathscr{P}(\mathbf{X})$ does not have the same cardinality as any subset of $\mathbf{X}$, we will assume that it has and show that this assumption leads to a contradiction (an indirect proof). Assume that $\#(\mathscr{P}(\mathbf{X})) = \#(\mathbf{S})$ and that $\mathbf{S} \subseteq \mathbf{X}$. This means that there is a 1-to-1 correspondence between $\mathbf{S}$ and $\mathscr{P}(\mathbf{X})$. Let $s$ in $\mathbf{S}$ correspond to $\mathbf{Y}_s$ in $\mathscr{P}(\mathbf{X})$. Since $\mathbf{Y}_s$ is a subset of $\mathbf{X}$, either $s$ is in $\mathbf{Y}_s$ or $s$ is not in $\mathbf{Y}_s$. Using this fact we can define a set $\mathbf{Y}_t$ which will be contained in $\mathbf{X}$. Let $\mathbf{Y}_t = \{s \text{ in } \mathbf{S} \mid s \text{ is not in } \mathbf{Y}_s\}$. (For examples of sets of the type $\mathbf{Y}_t$, see Exercises 3 to 5.) It is possible that $\mathbf{Y}_t$ is empty. In any case, since $\mathbf{Y}_t$ is in $\mathscr{P}(\mathbf{X})$, there is a $t$ in $\mathbf{S}$ such that $t \leftrightarrow \mathbf{Y}_t$ under the assumed 1-to-1 correspondence. Is $t$ in $\mathbf{Y}_t$? If $t$ is in $\mathbf{Y}_t$, then by the definition of $\mathbf{Y}_t$, $t$ is not in $\mathbf{Y}_t$. If $t$ is not in $\mathbf{Y}_t$, then by the definition of $\mathbf{Y}_t$, $t$ is a member of $\mathbf{Y}_t$. In either case, whether $t$ is in $\mathbf{Y}_t$ or whether $t$ is not in $\mathbf{Y}_t$, we reach a contradiction. This contradiction indicates that our original assumption that $\#(\mathscr{P}(\mathbf{X})) = \#(\mathbf{S})$, $\mathbf{S} \subseteq \mathbf{X}$, is false. Therefore, it is true that $\#(\mathbf{X}) < \#(\mathscr{P}(\mathbf{X}))$. ■

If we think for a minute about the statement of this theorem, $\#(\mathbf{X}) < \#(\mathscr{P}(\mathbf{X}))$, we will see that we have just proved a surprising fact. In the last section we divided sets into two classes—finite and infinite. We knew that finite sets come in various sizes (in fact, an infinity of sizes). Our first impulse may have been to think that all infinite sets are the same size. Cantor's theorem shows us that our first impression was not correct. The set $\mathbf{A} = \{1,2,3,\ldots\}$ is infinite, and so is $\mathscr{P}(\mathbf{A})$, but $\#(\mathbf{A}) < \#(\mathscr{P}(\mathbf{A}))$. Since $\mathscr{P}(\mathbf{A})$ is also a set, the theorem implies that $\#(\mathscr{P}(\mathbf{A})) < \#(\mathscr{P}(\mathscr{P}(\mathbf{A})))$, and we can continue in this way to form infinite sets each of which is larger than the preceding. In other words, there is an infinity of different-sized infinite sets.

Applied to the problem of arranging the twelve sets according to cardinality, from Cantor's theorem we know that $\#(\mathbf{A}) < \#(\mathbf{D})$ and $\#(\mathbf{H}) < \#(\mathbf{L})$. At the moment our list has the form

$$\mathbf{B} = \{P,Q,R,S\}$$
$$\mathbf{G} = \mathscr{P}(\mathbf{B}) = \{\mathbf{X} \mid \mathbf{X} \subseteq \mathbf{B}\}$$
$$\mathbf{C} = \{a,b,c,\ldots,y,z\}$$
$$\mathbf{F} = \mathscr{P}(\mathbf{C}) = \{\mathbf{X} \mid \mathbf{X} \subseteq \mathbf{C}\}$$
$$\mathbf{K} = \{\mathbf{X} \mid \mathbf{X} \text{ is a person alive in 1960}\}$$

$$\left\{\begin{array}{l} \mathbf{A} = \{1,2,3,\ldots\} \\ \mathbf{E} = \{10, 100, 1000, \ldots\} \\ \mathbf{I} = \{2,4,6,\ldots\} \\ \mathbf{J} = \{a/b \mid a \text{ and } b \text{ are natural numbers}\} \end{array}\right.$$

$$\mathbf{D} = \mathscr{P}(\mathbf{A}) = \{\mathbf{X} \mid \mathbf{X} \subseteq A\}$$

and although we know that $\mathbf{H} = \{n \mid n = 0.a_1a_2a_3\ldots$ and each $a$ is one of $0,1,\ldots,9\}$ must precede $\mathbf{L} = \mathscr{P}(\mathbf{H})$, we still cannot place $\mathbf{H}$ and $\mathbf{L}$ in the list since we do not know how $\mathscr{P}(\mathbf{A})$ and $\mathbf{H}$ compare. We will finally complete this problem in the next section.

## Exercises

1. Let $\mathbf{X} = \{1,2,3,4\}$. The power set of $\mathbf{X}$, $\mathscr{P}(\mathbf{X})$, was found in Exercise 1, Section 1. Show that there is a $\mathbf{Y}$ in $\mathscr{P}(\mathbf{X})$ such that there is a 1-to-1 correspondence between $\mathbf{X}$ and $\mathbf{Y}$.
2. Let $\mathbf{X} = \{1,2,3,\ldots\}$. Find $\mathbf{Y}$ in $\mathscr{P}(\mathbf{X})$ such that there is a 1-to-1 correspondence between $\mathbf{X}$ and $\mathbf{Y}$.
3. Let $\mathbf{S} = \{a,b,c\}$, $\mathbf{S} \subseteq \mathbf{X} = \{a,b,c,d\}$, and consider the correspondence

$$a \leftrightarrow \mathbf{Y}_a = \{b, c, d\}$$
$$b \leftrightarrow \mathbf{Y}_b = \{\ \ \}$$
$$c \leftrightarrow \mathbf{Y}_c = \{a, c\}.$$

Name all the members of the set $\mathbf{Y}_t = \{s \text{ in } \mathbf{S} \mid s \text{ is not in } \mathbf{Y}_s\}$. Show that $\mathbf{Y}_t$ is in $\mathscr{P}(\mathbf{X})$.

**4.** Let $\mathbf{S} = \{2,4,6,\ldots\}$, $\mathbf{S} \subseteq \mathbf{X} = \{1,2,3,4,\ldots\}$ and consider the correspondence

$$2 \leftrightarrow \mathbf{Y}_2 = \{1,2\}$$
$$4 \leftrightarrow \mathbf{Y}_4 = \{1,2,3,4\}$$
$$6 \leftrightarrow \mathbf{Y}_6 = \{1,2,3,4,5,6\}$$
$$\begin{matrix} \cdot & \cdot \\ \cdot & \cdot \\ \cdot & \cdot \end{matrix}$$

Name all the members of the set $\mathbf{Y}_t = \{s \text{ in } \mathbf{S} \mid s \text{ is not in } \mathbf{Y}_s\}$. Show that $\mathbf{Y}_t$ is in $\mathscr{P}(\mathbf{X})$.

**5.** Let $\mathbf{S} = \{1,3,5,7,\ldots\}$, $\mathbf{S} \subseteq \mathbf{X} = \{1,2,3,4,\ldots\}$ and consider the correspondence

$$1 \leftrightarrow \mathbf{Y}_1 = \{2\}$$
$$3 \leftrightarrow \mathbf{Y}_3 = \{2,4,6\}$$
$$5 \leftrightarrow \mathbf{Y}_5 = \{2,4,6,8,10\}$$
$$\begin{matrix} \cdot & \cdot \\ \cdot & \cdot \\ \cdot & \cdot \end{matrix}$$

Name all the members of the set $\mathbf{Y}_t = \{s \text{ in } \mathbf{S} \mid s \text{ is not in } \mathbf{Y}_s\}$. Show that $\mathbf{Y}_t$ is in $\mathscr{P}(\mathbf{X})$.

**6.** Let $\mathbf{A} = \{1,2,3,\ldots\}$. Name three members of each of the following sets: $\mathscr{P}(\mathbf{A})$, $\mathscr{P}(\mathscr{P}(\mathbf{A}))$, $\mathscr{P}(\mathscr{P}(\mathscr{P}(\mathbf{A})))$.

## 4. THE CARDINALITY OF THE SET OF REAL NUMBERS

In order to complete our arrangement of the list of sets we must now compare the cardinality of $\mathbf{D} = \mathscr{P}(\mathbf{A})$, the power set of the set of natural numbers, with the cardinality of $\mathbf{H} = \{n \mid n = 0.a_1a_2a_3\ldots$ and each $a$ is one of $0,1,\ldots,9\}$, the real numbers between 0 and 1 inclusive. It seems almost impossible to compare these two sets. Some of the members of $\mathscr{P}(\mathbf{A})$ are $\{1\}$, $\{1,3\}$, $\{10, 11, 73\}$, $\{2,4,7,11,\ldots\}$, while 0.271, 0.333..., 0.161616..., 0.0165 are members of $\mathbf{H}$. (If all of the $a$'s to the right of some position are 0, we write only the non-zero portion; for example, 0.27100... is written 0.271.)

***Find a similar property of the set $\mathscr{P}$(A) and the set H.***

One thing which is similar about **H** and $\mathscr{P}$(**A**) is that some of the decimal numbers in **H** contain a finite number of digits while some of the members of $\mathscr{P}$(**A**) are finite subsets of **A**. The remaining members of **H** are infinitely long decimal numbers, while the remaining members of $\mathscr{P}$(**A**) are infinite subsets of **A**. Perhaps this property is the clue to how to set up a correspondence between the two sets.

It is unfortunate that human beings have ten fingers. If we had only one finger on each hand the comparison we want to make would be easier. If human beings came equipped with a total of just two fingers, we would probably not base our numeration system on ten, but instead on two. Instead of the symbols 0, 1, 2, 3, 4, 5, 6, 7, 8, 9, we would use two symbols, 0 and 1, to form numerals. Instead of counting 1, 2, 3, 4, 5, 6, 7, 8, 9, 10, we would count 1, 10, 11, 100, 101, 110, 111, 1000, 1001, 1010. Instead of having a 1's, 10's, 100's, 1000's place, we would have a 1's, 2's, 4's, 8's place. This would be reflected in smaller numbers also. Instead of 0.1 meaning 1/10, it would mean 1/2 in base ten (or 1/10 in base two). Similarly, 0.01 instead of meaning 1/100 in base ten would mean 1/4 in base ten (or 1/100 in base two). We will suggest some practice with base-two notation in Exercises 1 to 4. The following list shows how some of the symbols translate from one system to the other.

| BASE TEN | BASE TWO |
|---|---|
| 100 | 1100100 |
| 40 | 101000 |
| 25 | 11001 |
| 17 | 10001 |
| 10 | 1010 |
| 5 | 101 |
| 2 | 10 |
| 1 = 0.999... | 1 = 0.111... |
| 0.625 | 0.101 |
| 0.5 | 0.1 |
| 0.333... | 0.010101... |
| 0.25 | 0.01 |
| 0.2 | 0.0011001100110011... |
| 0.0 | 0.0 |

***Show that each base-two symbol in the list above represents the same number as the corresponding base-ten symbol.***

The reason why we write $1 = 0.999\ldots$ in base ten and $1 = 0.111\ldots$ in base 2 will be explained in the exercises of Section 3, Chapter 7. In this chapter we will simply accept it as a known.

If we consider the set **U** of all numbers in base two which are between 0 and 1 (and including 0 and 1), we see a similarity with $\mathscr{P}(\mathbf{A})$, the power set of all natural numbers. The following correspondence would be natural:

$$
\begin{array}{ll}
\{\ \} & \leftrightarrow 0.0 \\
\{1\} & \leftrightarrow 0.1 \\
\{1,2\} & \leftrightarrow 0.11 \\
\{2,4,6\} & \leftrightarrow 0.010101 \\
\{2,4,6,\ldots\} & \leftrightarrow 0.010101\ldots \\
\{3,6,9,\ldots\} & \leftrightarrow 0.001001001\ldots \\
\{5,11\} & \leftrightarrow 0.00001000001 \\
\{1,2,3,\ldots\} & \leftrightarrow 0.111\ldots
\end{array}
$$

***Explain how the above correspondence was determined.***

The correspondence is established by thinking of the natural numbers as being listed in order according to size. Each natural number in the sets on the left corresponds to a position behind the decimal point (or maybe we should say "binary point"). The number 1 corresponds to the 1/2 position, 2 corresponds to the 1/4 position, 3 to the 1/8 position, and so on. A given subset will correspond to the base two symbol which has a 1 in the positions which correspond to all the elements in the subset and has 0 at all other positions. (If a base-two symbol terminates, as 0.1, we can think of it as the infinite symbol 0.1000. . . .)

This correspondence is 1-to-1, so $\#(\mathscr{P}(\mathbf{A})) = \#(\mathbf{U})$. If we lived in a world of two-fingered humans, we would say that the set of subsets of the natural numbers and the set of all real numbers between 0 and 1 are of the same cardinality.

Mathematics, of course, does not depend on the fact that we have ten fingers. Numbers are something quite independent of the symbols used to represent them. Numbers expressed in base ten or base two are still the same numbers. In fact, the number expressed in each of these bases establishes a 1-to-1 correspondence between the set of base-ten symbols in **H** and the set of base-two symbols in **U**. Since there is a 1-to-1 correspondence between $\mathscr{P}(\mathbf{A})$ and **U**, and a 1-to-1 correspondence between **U** and **H**, we can combine these correspondences to get a 1-to-1 correspondence between $\mathscr{P}(\mathbf{A})$ and **H**, $\#(\mathbf{H}) = \#(\mathscr{P}(\mathbf{A}))$.

***Give the final list of twelve sets arranged according to cardinality.***

$\mathbf{B} = \{P, Q, R, S\}$
$\mathbf{G} = \mathscr{P}(\mathbf{B})$
$\mathbf{C} = \{a, b, c, \ldots, y, z\}$
$\mathbf{F} = \mathscr{P}(\mathbf{C})$
$\mathbf{K} = \{\mathbf{X} \mid \mathbf{X}$ was a person alive in 1960$\}$

$$\left\{\begin{array}{l} \mathbf{A} = \{1,2,3,\ldots\} \\ \mathbf{E} = \{10, 100, 1000, \ldots\} \\ \mathbf{I} = \{2,4,6,\ldots\} \\ \mathbf{J} = \{a/b \mid a \text{ and } b \text{ are natural numbers}\} \end{array}\right.$$

$$\left\{\begin{array}{l} \mathbf{D} = \mathscr{P}(\mathbf{A}) \\ \mathbf{H} = \{n \mid n = 0.a_1a_2a_3\ldots \text{ and each } a \text{ is one of } 0,1,\ldots,9\} \end{array}\right.$$

$\mathbf{L} = \mathscr{P}(\mathbf{H})$

Now that our list is in order, there is one question which remains. How does the set **R** of all real numbers fit into this list? We have just shown that $\#(\mathbf{H}) = \#(\mathscr{P}(\mathbf{A}))$. If we consider $\mathbf{M} = \{x \mid x \text{ is in } \mathbf{H} \text{ and } x \neq 0 \text{ and } x \neq 1\}$, the set of all real numbers *strictly* between 0 and 1, we can also show that it is of the same cardinality as $\mathscr{P}(\mathbf{A})$ (and so the same cardinality as **H**). This will be done in Exercise 6. We can use **M** to determine the cardinality of *all* the real numbers.

***How does #(M) compare with the cardinality of all the real numbers?***

If **R** is the set of all real numbers, we know that $\mathbf{M} \subseteq \mathbf{R}$. However, we also know that it is possible for a proper subset of an infinite set to have the same cardinality as the entire set. In order to compare $\#(\mathbf{M})$ with $\#(\mathbf{R})$ we represent each of the sets on a number line. We pick a line and choose a

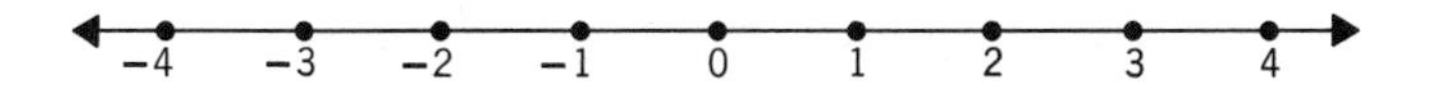

fixed point 0 on it. Then we decide on some unit of length and use it to lay off points to represent 1, 2, 3, . . . on one side of 0, and -1, -2, -3, . . . on the other side of 0. The set **M** is represented by all of the points between 0 and 1 (and not including 0 and 1). The real numbers **R** are represented by all of the points on the line. The problem of comparing #(**M**) with #(**R**) now becomes a question of comparing the number of points between 0 and 1 with the number of points on the entire line.

***How does the number of points between 0 and 1 compare with the number of points on the entire line?***

Once again, the correct answer is contrary to that our intuitition suggests. A 1-to-1 correspondence can be set up between the points strictly between 0 and 1 and the points on the entire line, so the cardinalities are the same. The diagram below indicates one way a 1-to-1 correspondence can be established.

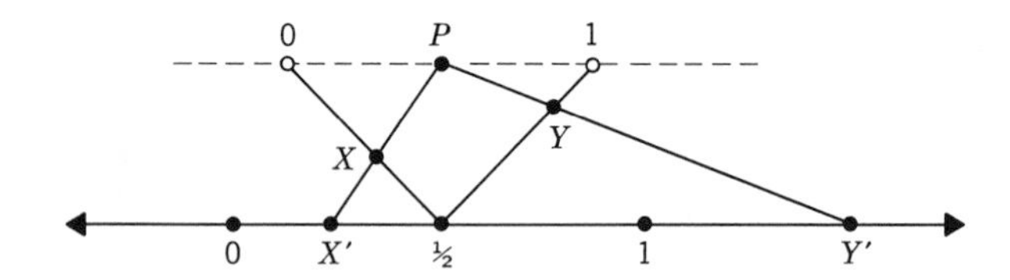


A copy of the segment between 0 and 1 is bent at 1/2 and placed at 1/2 on the number line so that each half makes the same angle with the number line. Point $P$ is chosen midway between 0 and 1 of the bent segment. A point $S$ on the segment corresponds to $S'$ on the number line if $S'$ is the intersection of the line $PS$ and the number line.

This geometrical correspondence shows us that $\#(\mathbf{M}) = \#(\mathbf{R})$. Exercise 6 will show that $\#(\mathscr{P}(\mathbf{A})) = \#(\mathbf{M})$ if **A** is the set of natural numbers. It follows that there are as many real numbers as there are subsets of the natural numbers, $\#(\mathbf{R}) = \#(\mathscr{P}(\mathbf{A}))$.

## Exercises

**1.** Count through the first twenty-five numbers in base two.

**2.** Change the following base-ten numerals to base two.

(a) 7 (c) 13 (e) 32 (g) 49 (i) 111

(b) 26 (d) 51 (f) 65 (h) 103 (j) 250

(If necessary, think of sticks arranged in bundles as little children do when they are learning base ten. For example, the base-ten numeral 11 is 1011 in base two because eleven sticks can be arranged into a bundle of eight, a bundle of two, and a bundle of one.)

**3.** Change the following base-two numerals to base ten.

| | | | |
|---|---|---|---|
| (a) 1111 | (c) 100 | (e) 1001 | (g) 10011000 |
| (b) 101010 | (d) 1100101 | (f) 10111 | (h) 1100110 |

**4.** Some base-ten decimals may be changed to base two by first changing the decimal to a fraction and then getting this fraction as a sum of $1/2$, $1/4$, $1/8$, . . . . For example, $0.75 = 3/4 = 1/2 + 1/4$ so 0.75 in base ten is the same number as 0.11 in base two. Change the following base-ten decimals to base two: 0.125, 0.0625, 0.5625, 0.625, 0.9375.

**5.** Use a geometrical argument to show that:

(a) If **M** is the set of all points between 0 and 1 on the number line (not including 0 and 1), then #(**M**) is equal to the cardinality of the positive real numbers.

(b) The number of points on the circumference of a circle of radius 1 inch is the same as the number of points on the sides of a square of side 2 inches.

(c) The number of points on the sides (all three) of an isosceles triangle of base 1 inch and altitude 1 inch is equal to the number of points on the sides (all four) of a square of side 1 inch.

***6.** Prove that if $\mathbf{M} = \{x \mid x \text{ is in } \mathbf{H} \text{ and } x \neq 0, x \neq 1\}$, that is the set of all real numbers strictly between 0 and 1, then $\#(\mathbf{M}) = \#(\mathscr{P}(\mathbf{A}))$. This can be done by using the known 1-to-1 correspondence between **H** and $\mathscr{P}(\mathbf{A})$ to show that $\#(\mathbf{M}) = \#(\mathbf{N})$ where **N** is all of the elements of $\mathscr{P}(\mathbf{A})$ except $\{1\}$ and $\{2\}$. It is then possible to show $\#(\mathbf{N}) = \#(\mathscr{P}(\mathbf{A}))$ by using the correspondence $\mathbf{X} \leftrightarrow \mathbf{X}$ if **X** is not of the form $\{r\}$ and $\mathbf{X} \leftrightarrow \{r-2\}$ if **X** is of the form $\{r\}$. Combining these two 1-to-1 correspondences we get $\#(\mathbf{M}) = \#(\mathscr{P}(\mathbf{A}))$.

## 5. CARDINAL NUMBERS

The discussion we have just had about different bases for the symbols (numerals) of numbers raises a question. What is a number? The *number three* is not the symbol 3 (in base ten) or 11 (in base two). The *number three* is merely represented by these symbols. The ancient Greeks said that *three* is a magnitude, it expresses the quantity of things. We can see three cars, or three crayons. We cannot see *three* itself. To the Greeks numbers were never far from the material things which they measured. We might say that for the Greeks, *three* was always an adjective which was used sometimes as a noun. Contemporary mathematicians always think of *three* as a noun, although they do not forget that it is an adjective also. Mathematicians today deal with *three* as if it were an object. For them, *three* has some reality of its own apart from any other objects it may describe.

At the end of the 19th century mathematicians began to try to define these numbers which they considered objects. Intuitively they thought of numbers which could be used to express the sizes of sets. These numbers were called *cardinal numbers.* An accurate definition of cardinal number is very difficult to formulate because of a contradiction which occurs in set theory unless certain precautions are taken. (This difficulty will be discussed in Section 7 of this chapter).

We have been saying that the cardinality of set **A** is equal to the cardinality of set **E** and writing $\#(\mathbf{A}) = \#(\mathbf{E})$. Intuitively we are saying to ourselves "**A** has a certain *number* of elements and **E** has the same *number* of elements." Instead of thinking of *same cardinality* as a relation between two sets, we are intuitively thinking of $\#(\mathbf{A})$ as assigning a number to **A**, and $\#(\mathbf{E})$ as assigning the same number to **E**.

In the finite cases as $\mathbf{B} = \{P, Q, R, S\}$ we think of $\#(\mathbf{B})$ as assigning the cardinal number *four* to **B**, and $\#(\mathscr{P}(\mathbf{B}))$ assigns the cardinal number *sixteen* to $\mathscr{P}(\mathbf{B})$. But what is the cardinal number *four?* We know that it is not the symbol 4. Even if we didn't know about this symbol we could think of the cardinal number of this set (as a small child does).

For example, the child might set up the following representatives for himself. He could arrange some blocks (or some other objects) as follows:

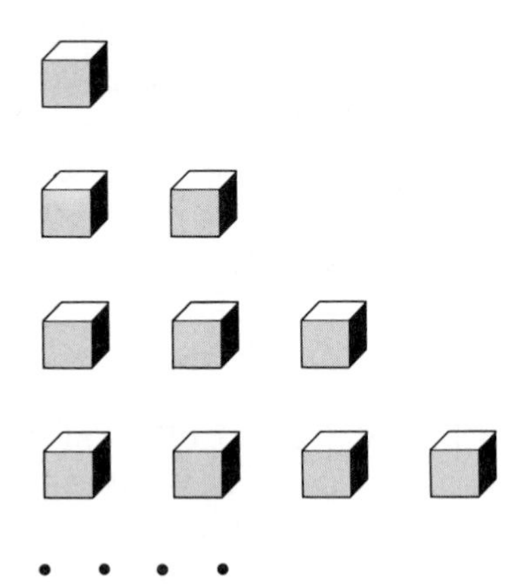

Now if we ask him, "How many cards do you have?" (and in fact he has four of them) he could say, "That many!" and point to

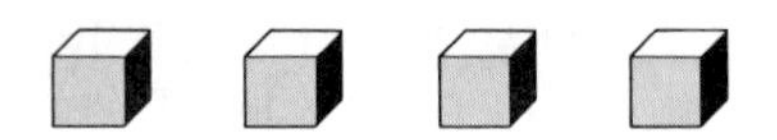

To this child, *four* is an object (this set of blocks) and not just a description of something else.

We can also make a number an object (rather than the magnitude of some other object) by setting up a system of representatives. Instead of blocks we can use sets of symbols. For convenience the symbols in these representative sets are the same ones we usually use for numbers.

$\{1\}$ will represent the cardinality of a set of one object;
$\{1,2\}$ will represent the cardinality of a set of two objects;
. . . .
$\{1,2,3,\ldots,n\}$ will represent the cardinality of a set of $n$ objects;
. . . .
$\{1,2,3,\ldots\}$ will represent the cardinality of any set which can be put in 1-to-1 correspondence with the set of natural numbers;
$\mathscr{P}(\{1,2,3,\ldots\})$ will represent the cardinality of any set which can be put in 1-to-1 correspondence with the real numbers;
$\mathscr{P}(\mathscr{P}(\{1,2,3,\ldots\}))$ will represent the cardinality of any set which can be put in 1-to-1 correspondence with the set of all subsets of the real numbers.

We can then define cardinal numbers to be all of these representative sets (and any other representatives we would pick for sets of cardinality different from those in our list). Whenever we ask the question, "How many objects are in this set?" our answer will be one of these representatives. For example,

$\#(\{a,b,c,d\}) = \{1,2,3,4\}$,
$\#(\{a,b,\ldots,y,z\}) = \{1,2,3,\ldots,26\}$, and
$\#(\{2,4,6,\ldots\}) = \{1,2,3,\ldots\}$.

Of course we do not want to have to write out these representative sets every time we use cardinal numbers, so it is natural to give these representative sets names. If we order the representative sets in the usual way, the finite cardinal numbers can be named by the last element in the representative set:

$\{1\} = \mathbf{1}$

$\{1,2\} = \mathbf{2}$

$\{1,2,3\} = \mathbf{3}$

. . .

$\{1,2,3,\ldots,n\} = \boldsymbol{n}$

We will use these oversized numerals to remind ourselves that they are the names of the cardinal numbers.

When we want to name the infinite cardinal numbers there is no last element to use, so we must invent some new symbols. One way to name them is to let

$$\{1,2,3,\ldots\} = \aleph_0$$

$$\mathscr{P}(\{1,2,3,\ldots\}) = \aleph_1$$

$$\mathscr{P}(\mathscr{P}(\{1,2,3,\ldots\})) = \aleph_2$$

where $\aleph$ is "aleph," the first letter of the Hebrew alphabet. Of course we can continue to pick representatives and names for larger infinite sets as

$$\mathscr{P}(\mathscr{P}(\mathscr{P}(\{1,2,3,\ldots\}))) = \aleph_3$$

$$\mathscr{P}(\mathscr{P}(\mathscr{P}(\mathscr{P}(\{1,2,3,\ldots\})))) = \aleph_4$$

$$\ldots$$

In this way we can find an infinity of infinite numbers!

## Exercises

**1.** What is the cardinal number of the following sets?
   (a) $\{a,b,c,d,e\}$
   (b) $\mathscr{P}(\{a,b,c,d,e\})$
   (c) $\{\mathbf{X} \mid \mathbf{X}$ is a state in the U.S.A.$\}$
   (d) $\{3,6,9,\ldots\}$
   (e) $\{a/b \mid a$ and $b$ are natural numbers$\}$
   (f) $\{n \mid n$ is a decimal number $0 < n < 1\}$

**2.** What is the difference between the natural numbers and the cardinal numbers?

**3.** Once the cardinal numbers are defined we can think of $\#(\mathbf{X})$ as assigning a cardinal number to $\mathbf{X}$; for example $\#(\{a,b\}) = \mathbf{2}$.
   (a) Let $\mathbf{X} = \{a,b,c\}$, $\mathbf{Y} = \{d,e\}$, $\mathbf{Z} = \mathbf{X} \cup \mathbf{Y}$. What are $\#(\mathbf{X})$, $\#(\mathbf{Y})$, $\#(\mathbf{Z})$? How do you think $\#(\mathbf{X}) + \#(\mathbf{Y})$ could be defined?
   (b) Let $\mathbf{X} = \{a,b,c\}$, $\mathbf{Y} = \{c,d\}$, $\mathbf{Z} = \mathbf{X} \cup \mathbf{Y}$. What are $\#(\mathbf{X})$, $\#(\mathbf{Y})$, $\#(\mathbf{Z})$? Does the definition of addition for cardinal numbers arrived at in (a) need any modification?

(c) Let $\mathbf{X} = \{0\}$, $\mathbf{Y} = \{1,2,3,\ldots\}$, $\mathbf{Z} = \mathbf{X} \cup \mathbf{Y}$. What are $\#(\mathbf{X})$, $\#(\mathbf{Y})$, $\#(\mathbf{Z})$? What is $1 + \aleph_0$?

(d) Let $\mathbf{X} = \{1,3,5,\ldots\}$, $\mathbf{Y} = \{2,4,6,\ldots\}$, and $\mathbf{Z} = \mathbf{X} \cup \mathbf{Y}$. What are $\#(\mathbf{X})$, $\#(\mathbf{Y})$, $\#(\mathbf{Z})$? What is $\aleph_0 + \aleph_0$?

## 6. WHAT HAVE WE LEARNED FROM CARDINAL NUMBERS?

There are three big lessons about mathematics to be learned from this chapter. First of all, this entire chapter grew out of the almost childish notion of a 1-to-1 correspondence and the question "What is a number?" This illustrates that mathematicians are frequently looking very hard at *simple* things.

A second lesson is that sometimes a few *simple assumptions can lead to an elegant and powerful theorem.* This was certainly the case when we considered Cantor's theorem about power sets. Using nothing but the definition of $<$ for cardinality of sets and the notions of power set and subset we were able to show that for any set $\mathbf{X}$, $\#(\mathbf{X}) < \#(\mathscr{P}(\mathbf{X}))$. The elegance of the proof in this case is due to the fact that no other tools were necessary (contrast this with the proof of the Theorem of Lagrange in Chapter 2), and yet the results are far reaching.

Cantor's theorem showed us immediately that infinite sets are not all of the same cardinality. This brings us to the third fact about mathematics that is illustrated by this chapter. Although intuition plays a big role in the discovery stages of mathematics, *often mathematics is unintuitive.* Our intuition would probably have said that all infinite sets had the same cardinality. Or perhaps we would have made the opposite mistake and said that $\{2,4,6,\ldots\}$ had fewer members than $\{1,2,3,\ldots\}$, since the first set is a part of the second. The problem is that we have had no real contact with infinite sets, so our intuition is undeveloped in this area and thus somewhat useless.

In our development of the cardinal numbers we found $\aleph_0 = \{1,2,3,\ldots\}$, $\aleph_1 = \mathscr{P}(\{1,2,3,\ldots\})$, and $\aleph_0 < \aleph_1$. We might wonder if there is an infinite set $\mathbf{Z}$ such that $\aleph_0 < \#(\mathbf{Z}) < \aleph_1$. To a certain extent this is still an open question. In 1963, however, it was answered for one commonly used axiom system of set theory (Zermelo's). Cohen proved that the assumption that there is an infinite set with cardinality between $\aleph_0$ and $\aleph_1$ can be added to this axiom system without causing any contradiction. Earlier Gödel had shown that the assumption that there is no infinite set with cardinality between $\aleph_0$ and $\aleph_1$ *(The Continuum Hypothesis)* is also consistent with the same axiom system. Both of these results show that it is possible to study different types of set theories, one with cardinals between $\aleph_0$ and $\aleph_1$ and one without.

Cohen's work is an example of good contemporary mathematics. He answered an important question which other mathematicians had been unable to answer; in the process he invented a technique which he has subsequently been able to apply to several other problems.

## Exercises

**READ ONE OR MORE OF THE FOLLOWING**

*Cohen, P. J., and Reuben Hersh, "Non-Cantorian Set Theory," *Scientific American,* Vol. 217, No. 6, December, 1967, pp. 104–116.

Hahn, Hans, "Infinity," in *The World of Mathematics,* Vol. 3, J. R. Newman, Simon and Schuster, New York, 1956, pp. 1593–1611.

Russell, Bertrand, "Definition of Number," in *The World of Mathematics,* Vol. 1, J. R. Newman, Simon and Schuster, New York, 1956, pp. 537–543.

Smullyan, Raymond M., "The Continuum Hypothesis," *The Mathematical Sciences,* Edited by COSRIMS, M.I.T. Press, Cambridge, Mass., 1969, pp. 254–260.

Spengler, Oswald, "Meaning of Numbers," in *The World of Mathematics,* Vol. 4, J. R. Newman, Simon and Schuster, New York, 1956, pp. 2315–2347.

### 7. WHAT IS SET THEORY?

Every branch of mathematics deals with some sets. The sets may be the set of natural numbers, the set of integers, a set of points, a set of lines, a set of rearrangements, a set of atomic statements, to mention just a few seen in this text. Thus, most systematic developments of mathematics presuppose first a development of set theory.

The first formal development of set theory was attempted by G. Frege beginning in 1879. He assumed axioms for set theory which seemed to express our intuitive idea of sets and then proceeded to deduce from these axioms two volumes of theorems. Among these intuitive ideas was the notion that every condition gives rise to a set (for example, the set of all people alive in 1960). Just as his second volume was going into print in 1902, he received a letter from Bertrand Russell which pointed out that his axioms lead to a contradiction.

Russell's reasoning was similar to the following. Some sets are members of themselves and some are not. For example, the set of all sets with more than one member is itself a set with more than one member. On the other hand, the set of natural numbers is not itself a natural number. Now let's consider **S**, the set of all sets which are not members of themselves, in symbols $\mathbf{S} = \{\mathbf{X} \mid \mathbf{X}$ is not a member of $\mathbf{X}\}$. Is **S** a member of itself? If **S** is a member of **S**, then by the description of **S**, **S** is not a member of **S**. This is a contradiction. But if we assume that **S** is not a member of **S**, then **S** satisfies the description, and so **S** is a member of **S**. Again a contradiction. Logically either **S** is in **S**, or **S** is not in **S**, yet both of these alternatives lead to a contradiction.

This contradiction is known as Russell's paradox. The discovery of this paradox showed that Frege's system was inconsistent. This must have been a great blow to Frege who had been laboring for over 20 years on his system. (Russell's paradox is just one of many intriguing paradoxes which arise in mathematics. See the article by Quine referenced at the end of this section.) The contradiction in Frege's system was due to the assumption that a set is determined every time a condition is specified; it is necessary to put some restrictions on this assumption. We will not go into the technical details, but will merely mention that this has been done by other mathematicians in a variety of ways.

The most frequently used alternative to Frege's system is Zermelo's set theory proposed in 1908. (Later improvements were added by Fraenkel and Skolem.) In Zermelo's system it is not possible to talk about "the set of all sets which are not elements of themselves," and so the Russell paradox does not arise. Although Zermelo avoided the paradox, his restrictions are not without problems. The notion of "the set of all sets" cannot be expressed in Zermelo's system either. It is this which causes much of the difficulty in arriving at an acceptable definition of cardinal number. Nevertheless, most parts of mathematics are now based on the foundation of Zermelo's set theory. Recently, the mathematicians working in the newly developed area of categorical algebra (see Section 8, Chapter 2) found that it is impossible to carry out their work in Zermelo's system. At the present time, mathematicians know that Zermelo's system is not the perfect answer for a foundation of mathematics and much work is being done to find a different approach.

The practical applications of set theory to the physical world arise in an indirect way. The work of Cantor, Frege, and Zermelo not only affected the foundations of mathematics, but led to a new attitude in mathematics. Their abstract approach to sets led to a more abstract approach to every branch of mathematics and more general approaches to classical problems in geometry, algebra, and analysis. New areas like topology, functional analysis, and combinatorial analysis developed. Each of these areas in turn has been rich in practical application. The giant step occurred with the beginning of abstract set theory.

## Exercises

### READ ONE OR MORE OF THE FOLLOWING

Breuer, Joseph, *Introduction to the Theory of Sets,* translated by Howard F. Fehr, Prentice-Hall, Englewood Cliffs, N. J., 1958.

Lieber, Lillian R., *Infinity,* Rinehart, New York, 1953, pp. 77–146.

*Quine, W. V., "Paradox," *Scientific American,* Vol. 206, No. 4, April, 1962, pp. 84-96.

## Test

1. True or False?
   (a) If $\mathbf{A} = \{a,b,c\}$, then $\{\mathbf{X} \mid \mathbf{X} \subseteq \mathbf{A}\}$ has 6 members.
   (b) $\#(\{1,2,3,\ldots\}) = \#(\{5,10,15,\ldots\})$.
   (c) $\mathscr{P}(\{1,2,3,\ldots,85\})$ has $85^2$ members.
   (d) For some $\mathbf{X}$, $\#(\mathbf{X}) = \#(\mathscr{P}(\mathbf{X}))$.
   (e) $\#(\{1,2,3,\ldots\}) < \#(\{1,2,3,\ldots,100\})$.
   (f) If $\mathbf{X} = \{x,y,z\}$, then $\{\{x\}, \{\ \ \}, \{y,z\}\}$ is a member of $\mathscr{P}(\mathbf{X})$.
   (g) There are as many points between 0 and 1 on the number line as there are points on the entire number line.
   (h) All infinite sets have the same cardinality.
   (i) A proper subset of a set may never have the same cardinality as the set itself.
   (j) $\#(\mathscr{P}(\{1,2,3,\ldots\})) = \aleph_2$

2. Rewrite each of the false statements in Question 1 so that it is a true statement.

3. Prove that $\#(\{x\}) < \#(\{r,s,t\})$ by using the definition of less than for cardinality.

4. Prove that $\#(\{1,2,3,\ldots\}) = \#(\{2,4,8,16,\ldots\})$.

5. Show that there is a 1-to-1 correspondence between the points on the circle and the points on the triangle in the sketch.

6. What outside reading have you done about set theory? What have you learned additional to the text from this reading?

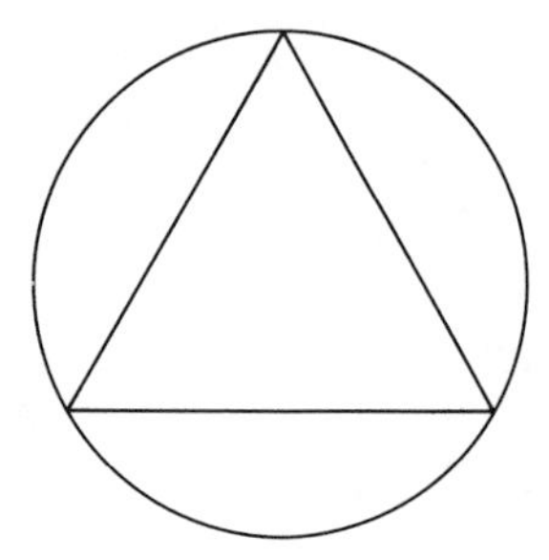

Axel Grosser-Monkmeyer

# 7

# ANALYSIS

## Approaching but Sometimes Never Getting There

**Analysis is the basic tool employed by the engineer who designed this bridge.**

The methods and achievements of analysis are among the great inventions of the human mind. Analysis is a prime example of mathematics as an art; at the same time, modern technology would have been impossible without it. Thus, in our attempt to understand the nature of mathematics, an excursion into analysis will be helpful.

In this chapter we will look at a fundamental tool of analysis—the notion of limit. Limit is the basic idea behind calculus, that part of analysis of which most of us have heard. When we finish we will not be able to read a calculus book, but we will have seen the key idea which is used in this part of mathematics.

## 1. FINITE SERIES

***Add the following column of numbers.***

$$\begin{array}{r} 3 \\ 8 \\ 2 \\ 7 \\ 5 \\ 1 \\ 6 \\ 9 \\ +\,5 \\ \hline \end{array}$$

There are several approaches to a simple calculation like this. Some people will start at the top and proceed down in order. Others will start at the bottom and go straight up. Still others will jump around and group the numbers in a convenient way. For example, someone could notice

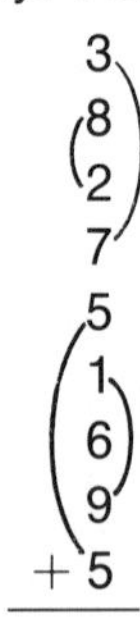

that the marked pairs add up to ten, and do the problem this way: $(3+7) + (8+2) + (1+9) + (5+5) + 6 = 46$.

The indicated sum of several numbers is called a *series.* We were considering the series $3+8+2+7+5+1+6+9+5$, which has the actual sum of 46. Each number in a series is called a *term.* Our series has nine terms; it is a *finite series* (see Chapter 6, Section 1 for a precise definition of finite). The person who rearranges and groups the terms of the series is using two properties of finite series which we know from arithmetic:

1. The terms of a finite series can be interchanged without changing the sum.
2. Parentheses may be added to (or removed from) a finite series without changing the sum.

In this chapter we will see how a basic calculation like this one can lead to some interesting generalizations. In this first section we will examine some short methods of finding the sum of a finite series. In the following sections we will try to generalize the notion of the sum of a finite series to the sum of a series which is not finite. We will see that different generalizations are possible. Along the way in the exercises we will ask whether parentheses can be added or removed in one of the generalizations. We will not go into the question of interchanging terms in the generalizations, but this question, too, has been studied by mathematicians.

The series we looked at above is made up of terms which follow no particular rule. Even in this case most of us look for a short way to find the actual sum. If the terms of a series do follow a rule, the challenge of finding the sum by some short method is more intriguing.

***What is the rule of formation in the following series?***

**(1)** $3+3+3+3+3+3$
**(2)** $0+5+0+5+0+5+0+5$
**(3)** $1+(-1)+1+(-1)+1+(-1)+1$
**(4)** $1+2+3+4+5+6+7+8+9$
**(5)** $1+3+5+7+9+11+13+15$

***Try to find a short way to calculate the actual sum of each series.***

In (1) the rule is that every term is 3. The sum can be found by multiplication; there are six terms and each is 3, so $3+3+3+3+3+3 = 6 \cdot 3 = 18$.

In (2) the odd-numbered terms (numbered from the left) are 0 and the even-numbered terms are 5. Since there are eight terms and half of them are 0 and half are 5, the sum can be found by dividing and multiplying, $(8/2)5 = 20$.

In (3) all the odd-numbered terms are 1 and the even-numbered terms are $-1$. If we group the terms as follows, $[1 + (-1)] + [1 + (-1)] + [1 + (-1)] + 1$, we see that the sum is $0 + 0 + 0 + 1 = 1$.

Series (4) follows the rule that the first term is 1, the second is 2, the third 3, and so on. One short method to find the sum is to notice that a series with a sum twice as large, $(1 + 2 + 3 + 4 + 5 + 6 + 7 + 8 + 9) + (1 + 2 + 3 + 4 + 5 + 6 + 7 + 8 + 9)$, can be rearranged and grouped as follows so that the sum is a multiple of 10: $(1 + 9) + (2 + 8) + (3 + 7) + (4 + 6) + (5 + 5) + (6 + 4) + (7 + 3) + (8 + 2) + (9 + 1) = 9 \cdot 10 = 90$. Therefore, the sum of the original series is half this number, or $90/2 = 45$.

Although this method may look like a trick which just happened to work in this case, it is more than that. It can be used to quickly find the sum of many finite series. Series (5), made up of the first eight successive odd natural numbers, is one example. A series with a sum twice as large as (5) is $(1 + 3 + 5 + 7 + 9 + 11 + 13 + 15) + (1 + 3 + 5 + 7 + 9 + 11 + 13 + 15) = (1 + 15) + (3 + 13) + (5 + 11) + (7 + 9) + (9 + 7) + (11 + 5) + (13 + 3) + (15 + 1) = 8 \cdot 16 = 128$. The sum of the original series is half this number, or $128/2 = 64$.

The methods we have used to find the sums of these five series can be applied to more general problems. Instead of considering the series $3 + 3 + 3 + 3 + 3 + 3$, we could consider a series with any finite number, say $N$, of 3s. The sum can still be found the same way, and is $N \cdot 3$. For example, $3 + 3 + 3 + \ldots + 3$ with 25 terms has sum $25 \cdot 3 = 75$.

***Find the sum of:***

***(A)*** $0 + 5 + 0 + 5 + \ldots$ ***with N terms***

***(B)*** $1 + (-1) + 1 + (-1) + \ldots$ ***with N terms***

***(C)*** $1 + 2 + 3 + 4 + \ldots + N$

***(D)*** $1 + 3 + 5 + 7 + \ldots$ ***with N terms***

If series (A) has an even number of terms, then the sum is $(N/2)5$. If $N$ is odd, it is $[(N-1)/2]5$. For example, if the series has 16 terms the sum is $(16/2)5 = 40$; if the series has 19 terms, the sum is $[(19 - 1)/2]5 = (18/2)5 = 45$.

If (B) has an even number of terms the sum is 0. If it has an odd number of terms the sum is 1.

To find the sum of (C) we use the same method as in (4) above. Twice the series is

$$[1 + 2 + 3 + \ldots + (N-1) + N] + [1 + 2 + 3 + \ldots + (N-1) + N] = (1 + N) + [2 + (N-1)] + \ldots + [(N-1) + 2] + (N + 1) = N \cdot (N + 1).$$

So the actual sum of (C) is $[N(N+1)]/2$. Using this formula we can quickly find the sum of any number of terms. For example, if $N=51$ then $1+2+3+\ldots+51=(51\cdot 52)/2=1326$.

To calculate the sum of (D) it would be helpful to know how to describe the $N$th term of the series in terms of $N$. The first term is 1, the second 3, the third 5, and so on and

$$\begin{aligned} 1 &= 2\cdot 1-1 \\ 3 &= 2\cdot 2-1 \\ 5 &= 2\cdot 3-1 \\ &\ldots \end{aligned}$$

so the $N$th term is $2N-1$. If the series has 30 terms, then the 30th term is $2\cdot 30-1=59$. Using the method of writing the series twice and then rearranging the terms, we obtain $[1+3+5+\ldots+(2N-3)+(2N-1)]+[1+3+5+\ldots+(2N-3)+(2N-1)]=[1+(2N-1)]+[3+(2N-3)]+\ldots+[(2N-3)+3]+[(2N-1)+1]=N(2N)=2N^2$. The sum of the original series is half as much so $1+3+5+\ldots+(2N-1)=N^2$.

## Exercises

**1.** What is the rule of formation of the following series? How many terms are in each series?

(a) $(-2)+2+(-2)+2+(-2)+2$
(b) $1+2+3+(-3)+(-2)+(-1)+1+2+3+(-3)+(-2)+(-1)+1$
(c) $3+3+3+3+3$
(d) $2+4+6+8+10+12$
(e) $3+6+9+12+15$
(f) $1+4+7+10+13+16+19$

**2.** Find the sum of each series in Exercise 1 by some short method.

**3.** For each of the series in Exercise 1, describe the $N$th term using $N$.

**4.** Find the sum of the following series in terms of $N$. (Notice that the rule of formation for each of the series is the same as in Exercise 1, so the $N$th term is as in Exercise 3.)

(a) $(-2)+2+(-2)+2+\ldots$ *with* $N$ terms
(b) $1+2+3+(-3)+(-2)+(-1)+1+2+3+(-3)+\ldots$ with $N$ terms
(c) $3+3+3+\ldots$ with $N$ terms
(d) $2+4+6+8+\ldots$ with $N$ terms
(e) $3+6+9+12+\ldots$ with $N$ terms
(f) $1+4+7+10+\ldots$ with $N$ terms

Check your answers for particular values of $N$.

**5.** Show that $1+1/2+1/4+1/8+1/16+1/32=2-1/32$. (Hint: Add 1/32 to the series and show that the sum of the new series is 2.) Show that $1+1/2+1/4+\ldots+1/2^R=2-1/2^R$ (notice that this equation becomes $1+1/2+1/4+\ldots+1/2^{N-1}=2-1/2^{N-1}$ when $N$ is the number of terms).

**6.** Find the sum of $1/(1\cdot 2)+1/(2\cdot 3)+1/(3\cdot 4)+1/(4\cdot 5)+1/(5\cdot 6)$. (Hint: Use the fact that $1/(1\cdot 2)=1/1-1/2$, $1/(2\cdot 3)=1/2-1/3$, and so on.) Also, find the sum of $1/(1\cdot 2)+1/(2\cdot 3)+1/(3\cdot 4)+\ldots+1/[N(N+1)]$. If $N=45$, what is the sum of this series?

***7.** Find the sum of $1/(1\cdot 2)+2/(1\cdot 2\cdot 3)+3/(1\cdot 2\cdot 3\cdot 4)+\ldots+N/[1\cdot 2\cdot 3\ldots N(N+1)]$. Notice that $1/(1\cdot 2)=1/1-1/(1\cdot 2)$, $2/(1\cdot 2\cdot 3)=1/(1\cdot 2)-1/(1\cdot 2\cdot 3)$, and so on. What is the sum of this series with six terms?

***8.** Find the sum of $1(1)+2(1\cdot 2)+3(1\cdot 2\cdot 3)+\ldots+N(1\cdot 2\cdot 3\ldots N)$. Notice that $1(1)=(1\cdot 2)-(1)$, $2(1\cdot 2)=(1\cdot 2\cdot 3)-(1\cdot 2)$, $3(1\cdot 2\cdot 3)=(1\cdot 2\cdot 3\cdot 4)-(1\cdot 2\cdot 3)$, and so on. If this series has seven terms, what is its sum?

***9.** What is the rule of formation in $4+8+21+52+65+96+\ldots$? What is the next term?

## 2. INFINITE SERIES

In Section 1 we noticed that the particular finite series $3+3+3+3+3+3$ could be generalized to a finite series with $N$ terms, $3+3+3+\ldots$ with $N$ terms. The sum of the first series is $6\cdot 3$, and the sum of the second is $N\cdot 3$. We now ask if a further generalization is possible. Can we talk about a series with an infinite number of terms? For example, $3+3+3+\ldots$ with as many terms as the natural numbers. (The natural numbers are the numbers we count with, 1, 2, 3, . . . .)

It is not difficult to describe such an infinite series. However, if we ask whether the infinite series $3+3+3+\ldots$ has a sum, there is a problem. We do not know what it means to add an infinite number of terms. The sum of a finite number of real numbers is always a real number. (The real numbers are all decimal numbers.) Is there a real number that we could assign as the "sum" of this infinite series which would be a natural extension of the sum of a finite series? If there is such a number, we feel that as we find the sums of longer and longer finite series consisting of the beginning terms of this infinite series we should be approaching the "sum."

***Show that by taking enough terms at the beginning of the infinite series $3+3+3+\ldots$ the sum can be made greater than 100, greater than 1000, greater than 1,000,000.***

If we consider the first 34 terms of the infinite series, the sum is $34 \cdot 3 = 102 > 100$. The sum of the first $N$ terms is called the $N$th *partial sum* of the series and is written $s_N$. In this example $s_1 = 3$, $s_2 = 6$, $s_3 = 9, \ldots, s_N = N \cdot 3$. The sum of the first 334 terms, $s_{334} = 334 \cdot 3 = 1002 > 1000$. The sum of the first 333,334 terms is $s_{333,334} = 333{,}334 \cdot 3 = 1{,}000{,}002 > 1{,}000{,}000$. We have no doubt that partial sums of this series can be found larger than any number we pick. It seems that for this particular infinite series no real number will be large enough to be the "sum" if "sum" is to be a natural extension of the sum of a finite series and the partial sums are to approach this number. We should be wondering at this point if this same difficulty will arise in every infinite series.

***Find the first ten partial sums of the following infinite series:***

***(A)*** $0 + 0 + 0 + \ldots$
***(B)*** $1/(1 \cdot 2) + 1/(2 \cdot 3) + 1/(3 \cdot 4) + \ldots$
***(C)*** $1 + 2 + 4 + 8 + \ldots$
***(D)*** $1 + (-1) + 1 + (-1) + \ldots$
***(E)*** $1 + 2 + 3 + (-3) + (-2) + (-1) + 1 + 2 + \ldots$
***(F)*** $1 + 1/2 + 1/4 + 1/8 + \ldots$
***(G)*** $1 + 1/2 + 1/3 + 1/4 + \ldots$
***(H)*** $(-2) + (-2) + (-2) + (-2) + \ldots$

The partial sums are as follows:

(A) 0, 0, 0, 0, 0, 0, 0, 0, 0, 0
(B) 1/2, 2/3, 3/4, 4/5, 5/6, 6/7, 7/8, 8/9, 9/10, 10/11
(C) 1, 3, 7, 15, 31, 63, 127, 255, 511, 1023
(D) 1, 0, 1, 0, 1, 0, 1, 0, 1, 0
(E) 1, 3, 6, 3, 1, 0, 1, 3, 6, 3
(F) 1, 3/2, 7/4, 15/8, 31/16, 63/32, 127/64, 255/128, 511/256, 1023/512
(G) 1, 3/2, 11/6, 25/12, 137/60, 147/60, 1089/420, 2283/840, 7129/2520, 7381/2520
(H) −2, −4, −6, −8, −10, −12, −14, −16, −18, −20

***Describe the behavior of these partial sums. Do you think that a "sum" could be defined for any of the above infinite series?***

In series (A) the partial sums are always 0. It seems reasonable to say that if a "sum" exists it should be 0.

In series (B) the succeeding partial sums seem to be getting closer to 1. We feel that the "sum" of this series should be 1, but it is not clear yet how to make the idea precise.

In series (C) the partial sums are growing larger at a rapid rate and they do not seem to be approaching any fixed real number. It does not seem possible to define a "sum" for this series.

In series (D) the partial sums oscillate between 0 and 1. Perhaps the "sum" should be 0, or maybe 1, but which one? Or maybe the "sum" could be the average 1/2, so that we wouldn't be partial to either value. This is all mere guessing, and it is not at all clear whether a "sum" can be defined for this series.

Series (E) is much like (D) except that the partial sums are 1, 3, 6, or 0, and the sums 1 and 3 appear twice as often as 6 and 0. There seems to be no simple way to assign a "sum" to this series.

The partial sums in series (F) are getting closer to 2 as we find the successive partial sums. In fact they do this in a very orderly way, $s_1 = 2 - 1$, $s_2 = 2 - 1/2$, $s_3 = 2 - 1/4$, . . . . We feel that a "sum" of 2 should be assigned to this series.

Series (G) is a problem. The partial sums are growing larger but rather slowly. None of the first ten partial sums reaches 3. The way in which the sums are increasing is not very clear. Whether they approach some definite number is hard to tell. We could calculate several more partial sums to help us make a better guess. However, since this is so time consuming we will wait until we have precise meaning for "sum" before passing judgment on this series.

Series (H) has partial sums which are all negative and which are growing larger negatively. They do not seem to approach any definite number and it does not seem possible to define a "sum" for this series.

In the eight examples we have looked at we can concentrate on two different kinds of behavior of the partial sums. In some cases the sums seem to get near to a definite number and stay there. For other series this does not happen.

Series (A), (B), and (F) all have partial sums which get near to a definite real number and stay near to it; (A) has partial sums that are always equal to 0 (so they are all near 0), (B) has partial sums which approach 1, and the partial sums of (F) approach 2. Series (C) and (H) have partial sums which grow without bound; in (C) they grow positively, in (H) negatively. In both cases the partial sums never stay near a definite number. Series (D) and (E) have partial sums which move back and forth over different values, so we cannot say that they stay near any one number. Series (G) is still a puzzle and at this point we cannot decide which category it belongs to.

What we need at this point is a precise definition of the notion of "getting near to a number and staying near." We will formulate this definition in Section 3. With it we will be able to extend the notion of sum of a finite series in some cases to the "sum" of an infinite series.

## Exercises

**1.** In each of the following series state the rule of formation:

(a) $2/4 + 3/4 + 4/4 + 5/4 + \dots$
(b) $3 + (-3) + 3 + (-3) + \dots$
(c) $1 + 1 + 1 + \dots$
(d) $3 + 0 + 3 + 0 + \dots$
(e) $1 + 1/3 + 1/9 + 1/27 \dots$
(f) $(-2) + (-2/5) + (-2/25) + \dots$
(g) $1 + 3/2 + 5/4 + 7/8 + \dots$

**2.** In each of the series in Exercise 1, find the next term.

**3.** In each of the series in Exercise 1, find the $N$th term.

**4.** In each of the series in Exercise 1, find the first eight partial sums.

**5.** In (a) to (e) of Exercise 1, find the $N$th partial sum.

**6.** Describe the behavior of the partial sums of each of the series in Exercise 1. Which ones have partial sums which seem to approach a definite real number? (Exercises 4 and 5 may help you do this exercise.)

### 3. CONVERGENT SERIES

In Section 2 we saw that some series have partial sums which get near a real number and stay there, others do not. In the first case we would like to say that the series converges to that real number and take the number to be the "sum" of the series. In the other case we want to say that the series does not converge, or does not have a "sum."

Before we introduce the precise definition of convergent series it might be helpful to think of the partial sums of a series geometrically. Let's consider a number line. On this line we label a certain point 0. We pick a unit

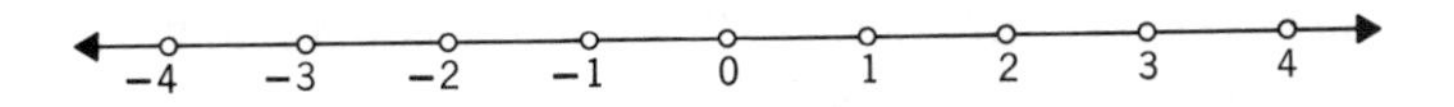

and label the points 1, 2, 3, ... at unit intervals on one side of 0. On the other side of 0 we label points −1, −2, −3, ... in a similar way. Every real number can then be related to a point on the line in a natural way by reference to the scale.

Since the partial sums of a series of real numbers are real, they can be represented as a set of points on the number line. The series 1 + 1/3 + 1/9 + 1/27 + ... has partial sums 1, 4/3, 13/9, 40/27, 121/81, .... They have the following geometrical representation in which the points are approaching 3/2.

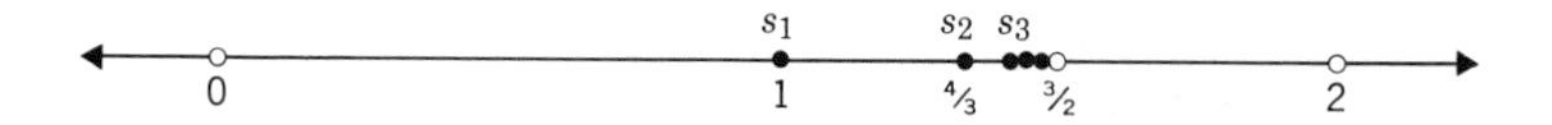


The series 1 + 1 + 1 + ... has partial sums represented in the following sketch:

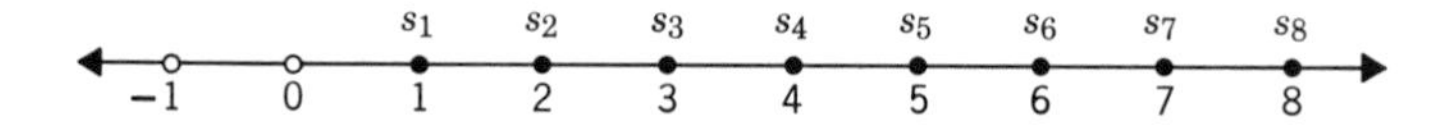


The partial sums of the series 1 + (−1) + 1 + (−1) + ... have the following geometrical representation.

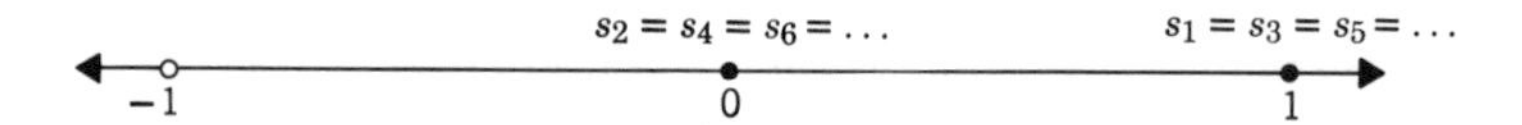


***Draw a geometrical representation of the partial sums of series (A) to (H) of Section 2.***

The geometrical representation of the partial sums of a series will help us define convergent series. Series

(A) $0 + 0 + 0 + \ldots$
(B) $1/(1 \cdot 2) + 1/(2 \cdot 3) + 1/(3 \cdot 4) + \ldots$
(F) $1 + 1/2 + 1/4 + 1/8 + \ldots$

of Section 2, are the type which we want to call convergent. Their geometrical representations are different from all of the others in one respect. In each case if we consider a very small portion of the number line around a certain point, we will find almost all of the partial sums in this portion, all except possibly a finite number.

In (A) any small portion of the number line about 0 contains literally all of the partial sums. This statement would not be true for any other point. The "sum" of this series is 0.

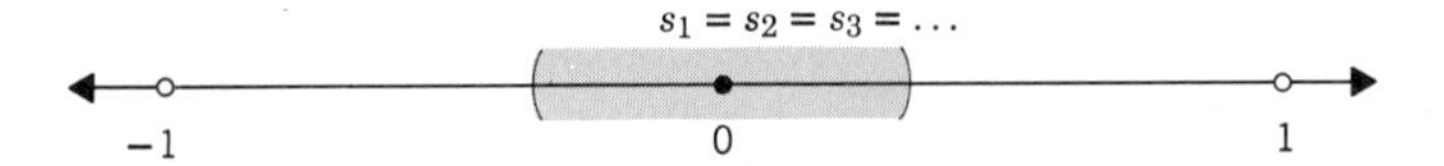


In (B) a small part of the number line about 1 leaves out very few of the partial sums. This statement is only true for 1. A small portion of the line about 1/2, for example 1/8 either side of 1/2, would leave out the infinite number of partial sums from $s_2$ on. The "sum" of this series is 1. Notice that the partial sums are approaching 1, but no partial sum actually equals 1.

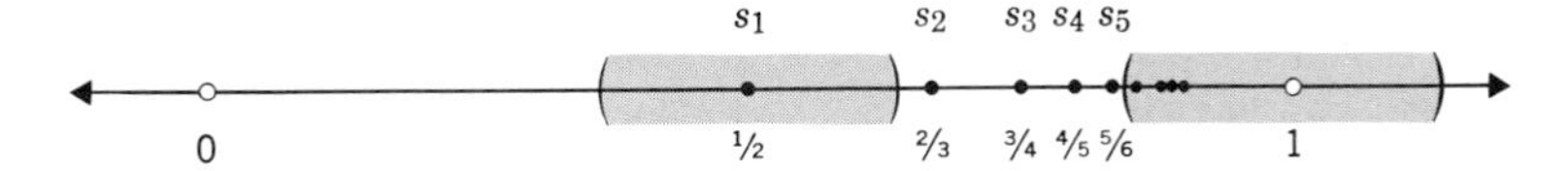


In (F) a small portion of the number line about 2 contains all but a finite number of the partial sums. The "sum" of this series is 2. Again in this case, no partial sum is equal to 2.

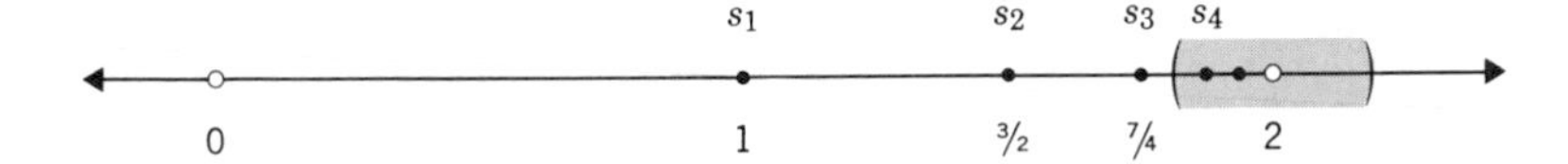


Let's consider two more series with a similar behavior:

(I) $(-1) + (-1/2) + (-1/4) + (-1/8) + \ldots$
(J) $1 + (-3/2) + 3/4 + (-3/8) + \ldots$

***Find the first eight partial sums of series (I) and (J). Draw their geometrical representations. Show that all but a finite number of the partial sums lie in a small portion of the number line about a certain point.***

We should notice that in all of these examples it does not matter how small a portion of the number line about the "sum" we consider. For example, in series (J) the partial sums are 1, −1/2, 1/4, −1/8, 1/16, −1/32, 1/64, −1/128, . . . so if we take a portion of the number line that measures 1/10 on each side of 0, we will miss only the first four partial sums, but will capture all those from $s_5$ on. That is, we will have $s_N$ in the small portion if $5 \leq N$.

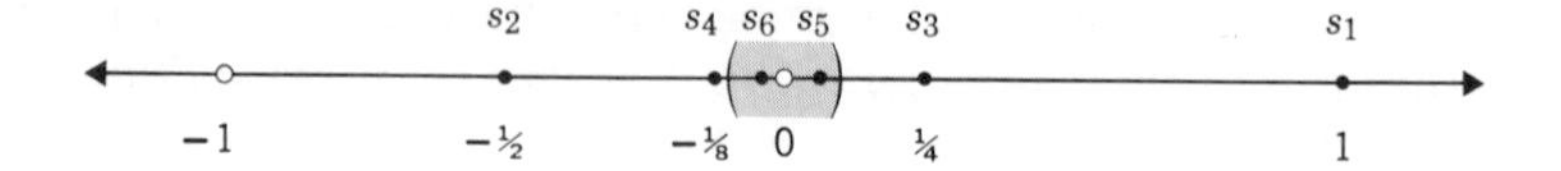


If we reduce the portion so that it measures 1/100 either side of 0 we will miss the first seven partial sums, but will get all of those from $s_8$ on. This is because for any $N$ with $8 \leq N$, $s_N$ differs from 0 by less than 1/100.

These pictures are very helpful to understand the notion of convergence. A working mathematician often draws pictures to help him gain insight and intuition about an idea. Pictures, however, are not mathematics. They can be deceiving. In order to reason logically about an idea we need a precise definition of the concept. We are now ready for a definition of convergence.

> DEFINITION: An infinite series, $a_1 + a_2 + a_3 + \ldots$, with partial sums $s_1 = a_1$, $s_2 = a_1 + a_2$, $s_3 = a_1 + a_2 + a_3, \ldots$ converges to the real number $S$ called the "sum" (or limit) of the series, if and only if, for any $1/10^K$ there is an $M$ such that $s_N$ and $S$ differ by less than $1/10^K$ for all $N \geq M$.

This definition is not as difficult as it appears at first reading. It merely says precisely in words what our pictures have been telling us. In our picture the small portion of the number line could be as small as we pleased; in our definition "for any $1/10^K$" indicates the size of this arbitrarily small portion. It measures $1/10^K$ on each side of $S$. (It might be helpful to think of $1/10^K$ in its decimal representation; it has a 1 in the $K$th place after the decimal point and 0's in all other places.) In the picture all but a finite number of the sums appear in the small portion; in words, "there is an $M$ such that $s_N$ and $S$ differ by less than $1/10^K$ for all $N \geq M$." The finitely many partial sums which are not in the small portion are $s_1, s_2, \ldots, s_{M-1}$.

Let's show now that the series in examples (A), (B), (F), (I), and (J) do indeed converge according to this definition. To show this we must indicate the "sum" $S$ and find a specific $M$ for every $K$ so that all partial sums from $s_M$ on are closer than $1/10^K$ to $S$.

In series (A), $0 + 0 + 0 + \ldots$, every partial sum is 0 and the difference between 0 and 0 is less than any $1/10^K$, so $S = 0$ and $M = 1$ for every $K$. Series (A) does have a "sum" of 0 according to the definition.

In series (B), $1/(1 \cdot 2) + 1/(2 \cdot 3) + 1/(3 \cdot 4) + \ldots$, it would be helpful to have a formula for $s_N$. This was found in Exercise 6 of Section 1, $s_N = 1 - 1/(N + 1)$. If we were asked to show that all but a finite number of the partial sums were within $1/10^{25}$ of 1 we could let $M = 10^{25}$. The partial sum $s_{10^{25}} = 1 - 1/(10^{25} + 1)$. This differs from 1 by $1 - s_{10^{25}} = 1/(10^{25} + 1) < 1/10^{25}$. Any partial sum after $s_{10^{25}}$ would be even closer to 0 since as the denominator increases, the fraction decreases. In a similar way for any $1/10^K$, $1 - s_{10^K} = 1/(10^K + 1) < 1/10^K$, so $S = 1$ and $M = 10^K$. The "sum" of (B) is 1 according to the definition of a convergent series.

Series (F), $1 + 1/2 + 1/4 + 1/8 + \dots$, has the $N$th partial sum $s_N = 2 - 1/2^{N-1}$ (see Exercise 5 of Section 1). For any $1/10^K$ we must show that there is an $M$ such that $2 - s_N = 1/2^{N-1} < 1/10^K$ for all $N \geq M$. This is true if we can make sure that the denominator $2^{N-1} > 10^K$. One way is to notice that $2^4 > 10$, $(2^4)^2 > 10^2$, $(2^4)^3 > 10^3$, ..., $(2^4)^K > 10^K$. Since $(2^4)^K = 2^{4K}$ we can let $M = 4K + 1$, then $2 - s_M = 1/2^{M-1} = 1/2^{(4K+1)-1} = 1/2^{4K} < 1/10^K$ and for any number $N$ greater than $M$ the difference is even smaller. So in series (F) $S = 2$ and for any $K$, $M = 4K + 1$.

Series (I), $(-1) + (-1/2) + (-1/4) + (-1/8) + \dots$, is like (F) except that all of the partial sums are negative. This means that for (I) the "sum" is $S = -2$. Notice that for any $K$, $M$ still has value $4K + 1$. (In this case, however, to find how much $s_N$ differs from $S$ we must find $s_N - S$ to obtain a positive number.)

In series (J), $1 + (-3/2) + 3/4 + (-3/8) + \dots$, the odd partial sums are $1/2^{N-1}$ and the even ones are $-1/2^{N-1}$. The odd partial sums differ from 0 by $s_N - 0 = 1/2^{N-1}$ and the even partial sums differ from 0 by $0 - s_N = 1/2^{N-1}$. As in the last two series, for any $K$ we can take $M = 4K + 1$ and $1/2^{M-1} = 1/2^{(4K+1)-1} = 1/2^{4K} < 1/10^K$. For any $N \geq M$ the denominator is even smaller, so the "sum" of series (J) is 0.

For series (A), (B), (F), (I), and (J), and any other series which converges we may now speak of the "sum" (or limit) of the series. The "sum" is the number $S$ to which the series converges. Notice that in some of these cases (such as (B), (F), (I), and (J)) no partial sum actually equals the "sum." It is possible for the partial sums to approach the "sum" without ever getting there. There are, of course, series in which some partial sums actually equal the "sum." Series (A) is one; others will be mentioned in the exercises.

We might be wondering at this point why we go through all this work to define convergence. It seems that we knew just as much by drawing pictures. Series (G), $1 + 1/2 + 1/3 + 1/4 + \dots$, is a good illustration of the need for a precise approach. The geometrical representation of the first few partial sums is inconclusive. The series may be convergent or one

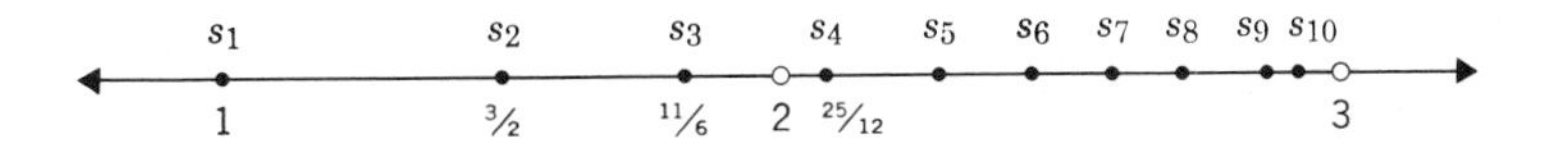


which does not converge. The truth is that this series does not have a "sum," and in fact the partial sums grow as large as we want. In the next section we will prove that series (G) and the other remaining examples (C), (D), (E), and (H) do not have "sums."

It is unfortunate that we have not been able to generalize the notion of finite sum to *all* infinite series. Maybe we could define "sum" a different way so that a larger class of infinite series will have a "sum." This is possible and we will look at it in Section 5.

## Exercises

**1.** Draw the geometrical representation of the first eight partial sums of the following series:

(a) $2 + 0 + 2 + 0 + \ldots$
(b) $[-1/(1 \cdot 2)] + [-1/(2 \cdot 3)] + [-1/(3 \cdot 4)] + \ldots$
(c) $2/4 + 3/4 + 4/4 + 5/4 + \ldots$
(d) $1/10 + 1/10^2 + 1/10^3 + \ldots$
(e) $(-3) + 3 + (-3) + 3 + \ldots$
(f) $1 + (-1/2) + 1/2 + (-1/4) + 1/4 + (-1/8) + 1/8 + \ldots$

**2.** Show that the "sum" $[-1/(1 \cdot 2)] + [-1/(2 \cdot 3)] + [-1/(3 \cdot 4)] + \ldots$ is $-1$. (Hint: Notice that this series is similar to series (B) in this section.)

**3.** For the series in (d) of Exercise 1:

(a) Find the $N$th term;
(b) Show that the $N$th partial sum is $1/9 - 1/(9 \cdot 10^N)$;
(c) Prove that this series converges to 1/9;
(d) Write the series as a decimal number.

**4.** In series (f) of Exercise 1 what is the $N$th partial sum if $N$ is even? If $N$ is odd? Show that the "sum" of this series is 1. (Notice that this is a series in which some of the partial sums actually equal the "sum.")

**5.** A finite series as $4 + 3 + 2 + 1$ can be thought of as an infinite series in which all of the terms after a certain one are 0. The finite series just mentioned can be identified with the infinite series $4 + 3 + 2 + 1 + 0 + 0 + \ldots$. Use the definition of convergence to show that the "sum" of this infinite series is the same as the ordinary sum of the associated finite series. (In the case of this infinite series notice that the partial sums approach the "sum" and then actually equal it.)

**6.** The repeating decimal $0.999\ldots$ can be thought of as an infinite series, $9/10 + 9/10^2 + 9/10^3 + \ldots$. Show that this infinite series has a "sum" of 1. (This is why when dealing with decimals we can identify $0.999\ldots$ with 1.)

***7.** The number $0.111\ldots$ in base two can be thought of as the infinite series $1/2 + 1/4 + 1/8 + \ldots$ in base 10. Why should this number be identified with the number 1?

***8.** Show that the repeating decimal 0.333... should be identified with 1/3, that is, the "sum" of $3/10 + 3/10^2 + 3/10^3 + \ldots$ is 1/3.

**9.** The terminating decimal 0.125 can be thought of as an infinite decimal with 0s in all places from the thousandths place on, 0.125000.... This decimal is then the infinite series $1/10 + 2/10^2 + 5/10^3 + 0/10^4 + 0/10^5 + \ldots$. Show that the "sum" of this infinite series is 1/8.

## 4. SERIES WHICH DO NOT CONVERGE

In this section we will show that series

$$\begin{array}{ll} \text{(C)} & 1 + 2 + 4 + 8 + \ldots \\ \text{(D)} & 1 + (-1) + 1 + (-1) + \ldots \\ \text{(E)} & 1 + 2 + 3 + (-3) + (-2) + (-1) + 1 + 2 + \ldots \\ \text{(G)} & 1 + 1/2 + 1/3 + 1/4 + \ldots \\ \text{(H)} & (-2) + (-2) + (-2) + \ldots \end{array}$$

do not satisfy the definition of convergence and therefore do not have "sums."

We can treat series (C), (G), and (H) together because in each case the partial sums grow without bound. In (C) and (G) they grow positively, in (H) they grow negatively. Given any positive number $T$ we can show that after a certain partial sum all of the partial sums of (C) and (G) are greater than $T$. Given any negative number $-U$, we can show that after a certain partial sum all of the partial sums of (H) are less than $-U$. Let us show this now.

In (C) it is easy to see this. If $T = 1{,}000{,}000$, then certainly every partial sum from $s_{1{,}000{,}001}$ on is greater than $T$ because every term in the partial sum is 1 or greater than 1. (Of course this happens before $s_{1{,}000{,}001}$, but we are not concerned with precisely at which partial sum it occurs, but that in fact it does occur some place.)

***For any* T*, show that there is a natural number* M *such that all partial sums* $s_N$ *of* $1 + 2 + 4 + 8 + \ldots$ *are greater than* T *if* $N \geq M$.**

The general case is similar to the specific one above for $T = 1{,}000{,}000$. For any $T$, $s_N > T$ for all $N \geq M = T + 1$ because every term in $s_N$ is 1 or greater. The partial sums for this series grow larger and larger without bound.

***Show that for any* T *there is a natural number* M *such that all partial sums* $s_N$ *of* $1 + 1/2 + 1/3 + 1/4 + \dots$ *are greater than* T *if* $N \geq M$.**

If we get an estimate of some of the partial sums we see that

$$
\begin{aligned}
s_2 &= s_{2^1} > 1/2\\
s_4 &= s_{2^2} > 2/2\\
s_8 &= s_{2^3} > 3/2\\
s_{16} &= s_{2^4} > 4/2\\
&\dots\dots\\
s_{2^T} &> T/2\\
&\dots\dots\dots\\
s_{2^{2T}} &> T
\end{aligned}
$$

because in

$$\underbrace{1 + 1/2} + \underbrace{1/3 + 1/4} + \underbrace{1/5 + 1/6 + 1/7 + 1/8} + \underbrace{1/9 + 1/10 + \dots + 1/16} + \dots$$

each of the indicated portions is greater than 1/2. This means that the partial sums continue to get larger without bound. If we want a partial sum greater than 100, then $s_{2^{200}}$ certainly would be and also any partial sum after that since the partial sums are increasing. In general, for any $T$ all partial sums $s_M$ with $N \geq M = 2^{2T}$ are greater than $T$. So this series instead of resembling a convergent series, is more like $1+2+4+8+\dots$. The only difference is that the rate of growth is slower.

***Show that for any* $-U$ *with* $U > 0$ *there is a natural number* N *such that the partial sums* $s_N$ *of* $(-2) + (-2) + (-2) + \dots$ *with* $N \geq M$ *are all less than* $-U$.**

Since each term of $s_N$ is $-2$, $s_N < -N$. For example, $s_{100} = -200 < -100$. In general for any negative number $-U$, $s_N < -U$ for all $N \geq M = U$.

In all three of these cases the series do not converge because if we take any portion of the number line of length $1/10^K$ about any point $S$, the partial sums instead of eventually converging into that portion, from some $M$ on are all outside of the portion. In (C)

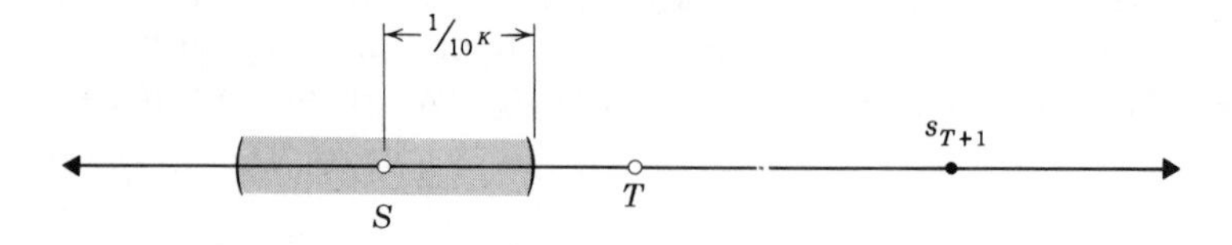

if we take $T$ to be an integer such that $S + 1/10^K \leq T$, then $S_N > T$ for all $N \geq T + 1$. In (G), $s_N > T$ for all $N \geq 2^{2T}$.

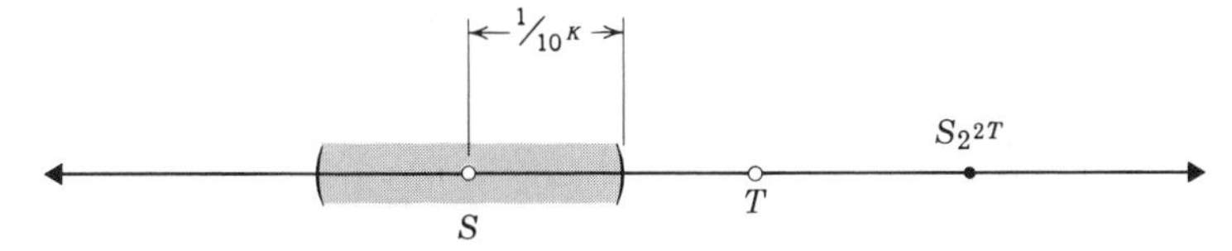


In (H), if we let $-U$ be a negative integer such that $-U \leq S - 1/10^K$

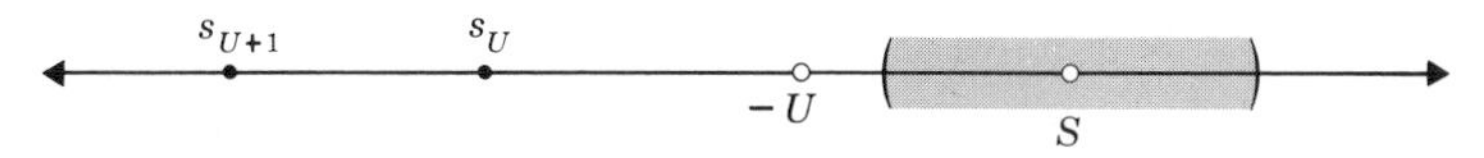


then $s_N < -U$ for all $N \geq U$.

Series (D) and (E) are a little different than these other three, but they still do not converge according to our definition.

In (D) the partial sums are 1, 0, 1, 0, . . . . The series does not converge to 1 because a portion of the number line of length 1/10 about 1 will leave out an infinite number of partial sums, namely $s_2 = s_4 = s_6 = \ldots$. Similarly it does not converge to 0. No other real number $S$ could be the sum either since we could take a small portion of length $1/10^K$ on either side of $S$ and none (or possibly only half) of the partial sums would be in this portion.

***Show that series (E) does not converge to any real number.***

Our proof should be similar to the one given above for series (D).

It is a little discouraging that our definition of "sum" of an infinite series does not apply to so many series. In Section 4 we will try to improve it so that more infinite series will have "sums."

## Exercises

**1.** Prove that the following series do not converge:

(a) $1 + 2 + 3 + 4 + 5 + \ldots$
(b) $1 + 1 + 1 + \ldots$
(c) $(-1) + (-1) + (-1) + \ldots$
(d) $1/4 + 2/4 + 3/4 + 4/4 + 5/4 + \ldots$
(e) $(-3) + 3 + (-3) + 3 + (-3) + \ldots$

(f) $1/2+(-1/2)+1+(-1)+1/2+(-1/2)+\dots$
(g) $(-1/8)+1+(-15/16)+1+(-31/32)+1+(-63/64)+\dots$

**2.** Make up a series which does not converge because its partial sums grow large positively without bound.

**3.** Make up a series which does not converge because its partial sums grow negatively without bound.

**4.** Make up a series which does not converge because it has partial sums which oscillate between two numbers.

**5.** Make up a series which does not converge because it has partial sums which move over three values.

**6.** Make up a series which has partial sums which approach two different values, but the partial sums never equal either one of them (as the series in part (g) of Exercise 1).

**7.** In the following questions the addition of parentheses to series is considered.
(a) The series $1+(-1)+1+(-1)+\dots$ does not converge. Consider a similar series with parentheses added, $[1+(-1)]+[1+(-1)]+\dots$. Does it converge?
(b) The series $1+1+1+1+\dots$ does not converge. Consider a similar series with parentheses added $(1+1+1)+(1+1+1)+\dots$. Does it converge?
(c) The series $0+0+0+\dots$ converges. Does $0+(0+0)+(0+0+0)+\dots$ converge? If so, does it converge to the same "sum"?
(d) The series $1+(-3/2)+3/4+(-3/8)+\dots$ converges. Does $[1+(-3/2)]+[3/4+(-3/8)]+\dots$ converge? If so, does it converge to the same "sum"?
(e) Make a conjecture about adding parentheses to infinite series.

**8.** In the following questions removing parentheses from infinite series is considered.
(a) Does the series $[2+(-2)]+[2+(-2)]+\dots$ converge? If the parentheses (square brackets) are removed, does it converge?
(b) Does the series $(1/2+1/2)+(1/4+1/4)+(1/8+1/8)+\dots$ converge? If the parentheses are removed does it seem to converge?
(c) Does the series $(0+0+0)+(0+0+0)+\dots$ converge? If the parentheses are removed does it converge?
(d) Does the series $[1+(-1)]+[1/2+(-1/2)]+[1/4+(-1/4)]+\dots$ converge? If the parentheses are removed does it seem to converge?
(e) Does $[1/2+(-5/8)]+[3/4+(-13/16)]+[7/8+(-29/32)]+\dots$ converge? If the parentheses are removed does it converge?
(f) Does $(1/10+1/10+1/10)+(1/10^2+1/10^2+1/10^2)+\dots$ converge? If the parentheses are removed does it converge?

**9.** What is wrong with the following reasoning?
$1+(-1)+1+(-1)+\dots=1+[(-1)+1]+[(-1)+1]+\dots=1$ and $1+(-1)+1+(-1)+\dots=[1+(-1)]+[1+(-1)]+\dots=0$, therefore $1=0$.

## 5. ANOTHER TYPE OF CONVERGENCE

We have seen in Section 4 that the partial sums $s_1, s_2, s_3, \ldots$ of the infinite series $1 + (-1) + 1 + (-1) + \ldots$ oscillate between 1 and 0 and so the series does not have a "sum."

There is no reason why "sum" must be defined as it was in Section 3. Mathematicians are free to define things any way they please—within certain limits. They cannot introduce a definition or an axiom which causes a contradiction (the Russell paradox in Section 7, Chapter 6, is an example of this). In practice it is often difficult to check this. Second, they always must be sure that the thing they are defining exists.

***What is wrong with the following? Define R to be a natural number such that $1 < R$ and $R < 2$. Then $1 + 1 = 2 < R + 1$, $R + 1 < 2 + 1 = 3$, and $R + 1$ is a natural number. This means that there is a natural number between 2 and 3. Similarly, $3 < R + 2 < 4$, $4 < R + 3 < 5, \ldots$, and $R + 2$, $R + 3, \ldots$ are all natural numbers. Therefore, there is a natural number between any two consecutive natural numbers.***

The mistake is that there is no natural number between 1 and 2. The definition of $R$ describes a nonexistent object. Any reasoning about such an object is meaningless. Whenever a mathematician introduces a definition he must check that there is at least one object that satisfies his definition.

There is a third requirement of a definition which is more difficult to describe. Although the mathematician has great freedom in introducing definitions, his decision is guided by another element besides being free of contradiction and nonexistence. Correct logic is not the only criterion of good mathematics. He also considers whether the definition describes something significant and useful within mathematics. For example, someone might look at the primes, 2, 3, 5, 7, 11, 13, 17, . . . and divide them into two sets by considering those which are first, third, fifth, and so on in the list and those which are in the even positions. He might say, "Let's define a first-order prime to be one of 2, 5, 11, 17, . . . and a second-order prime to be one of 3, 7, 13, . . . ." This definition will not cause a contradiction since it merely gives a name to things which we already have. First- and second-order primes as defined exist. However, at this time no one has discovered anything of interest about these first- and second-order primes, so to introduce the definition is useless.

Let's return now to our problem of defining a "sum" of an infinite series. The definition of "sum" in Section 3 has turned out to exclude many series. We could try to define it another way—remembering of course the criteria for a good definition.

One approach would be to take the average of the partial sums. For example, the partial sums of $1 + (-1) + 1 + (-1) + \ldots$ are 1, 0, 1, 0, .... The average of $s_1$ is $t_1 = s_1/1 = 1/1 = 1$; the average of $s_1$ and $s_2$ is $t_2 = (s_1 + s_2)/2 = (1 + 0)/2 = 1/2$; the average of $s_1$, $s_2$, and $s_3$ is $t_3 = (s_1 + s_2 + s_3)/3 = (1 + 0 + 1)/3 = 2/3$. In general the average of the first $N$ partial sums is $t_N = (s_1 + s_2 + \ldots + s_N)/N$.

---

***Find $t_1$, $1_2$, ..., $t_{10}$ for the series $1 + (-1) + 1 + (-1) + \ldots$ and draw a geometrical representation of these averages.***

---

For the infinite series $1 + (-1) + 1 + (-1) + \ldots$ the first ten averages of the partial sums are 1, 1/2, 2/3, 2/4, 3/5, 3/6, 4/7, 4/8, 5/9, and 5/10. The averages $t_N$ for even $N$ are all equal to 1/2. The other averages are approaching 1/2. If the averages approach a number we will define the "sum according to the first arithmetic mean" to be the number which the averages approach. The formal definition looks as follows:

> DEFINITION: A series $a_1 + a_2 + a_3 + \ldots$ with partial sums $s_1, s_2, s_3, \ldots$ and averages $t_N = (s_1 + s_2 + \ldots + s_N)/N$ converges according to the first arithmetic mean to the real number $T$, if and only if, for any $1/10^K$ there is a natural number $M$ such that $t_N$ and $T$ differ by less than $1/10^K$ for all $N \geq M$.

This definition does describe something which exists. The infinite series $1 + (-1) + 1 + (-1) + \ldots$ has a "sum according to the first arithmetic mean" as shown above. We will not go into the question of whether or not this definition could cause a contradiction. We feel that it will not since finding an average is an accepted operation in arithmetic and the notion of convergence is the basic idea in analysis.

In order to see that the definition is useful, we should compare it with our original definition of "sum" in Section 3. Our first definition of "sum" was a generalization of the sum of a finite series. Any finite series can be thought of as an infinite series with all 0's after a certain term. The sum of a finite series and the "sum" of its related infinite series are the same, so our first generalization preserved all of the original sums. We wonder now if the "sum according to the first arithmetic mean" will preserve all the "sums" we had before.

---

***Find $t_1$, $t_2$, ..., $t_8$ for the infinite series $1 + 1/2 + 1/4 + 1/8 \ldots$. Do they seem to approach the "sum" of this series as found in Section 3?***

---

The "sum" of this series as found in Section 3 is 2. The averages are 1, 5/4, 17/12, 49/32, 129/80, 321/192, 769/448, 1793/1024. The averages seem to be approaching 2, but we cannot be certain from these few specific cases. Rather than proving that these averages do approach 2, it would be more profitable to show that every infinite series which has a "sum" according to our original definition also has a "sum according to the first arithmetic mean" and that these numbers are the same.

Before we state and prove this theorem, we will introduce an idea that will streamline the proof. Because in series we deal with both positive and negative terms, we have had to be careful when we found the difference between certain numbers. In order to have a positive difference, we sometimes subtracted in one order, sometimes in the other. We can avoid this by using the notion of *absolute value.* The absolute value of a real number $x$, written $|x|$, is $x$ if $x \geq 0$ and is $-x$ if $x < 0$. For example $|3| = 3$, $|0| = 0$, and $|-11| = 11$.

There are two useful properties of absolute value which are easy to see.

---

| ***Compare*** | $\boldsymbol{|3/2 + 5|}$ | ***and*** | $\boldsymbol{|3/2| + |5|}$, |
|---|---|---|---|
| | $\boldsymbol{|0.4 + 0|}$ | ***and*** | $\boldsymbol{|0.4| + |0|}$, |
| | $\boldsymbol{|-3 + 2|}$ | ***and*** | $\boldsymbol{|-3| + |2|}$, |
| | $\boldsymbol{|-3 + (-5)|}$ | ***and*** | $\boldsymbol{|-3| + |-5|}$, |
| | $\boldsymbol{|5 + (-2)|}$ | ***and*** | $\boldsymbol{|5| + |-2|}$. |

***Make a conjecture.***

---

In each case the first quantity is equal or less than the second. We conjecture that for all real numbers $x$ and $y$, $|x + y| \leq |x| + |y|$. This is a correct conjecture. We will not stop to prove it, but we will feel free to use it and to extend it to any finite number of terms; for example, $|a + b + c + d| \leq |a| + |b| + |c| + |d|$.

---

| ***Compare*** | $\boldsymbol{|3 \cdot 5|}$ | ***and*** | $\boldsymbol{3|5|}$, |
|---|---|---|---|
| | $\boldsymbol{|7(-10)|}$ | ***and*** | $\boldsymbol{7|-10|}$, |
| | $\boldsymbol{|8 \cdot 0|}$ | ***and*** | $\boldsymbol{8|0|}$, |
| | $\boldsymbol{|0(-3)|}$ | ***and*** | $\boldsymbol{0|-3|}$, |
| | $\boldsymbol{|(1/2)7|}$ | ***and*** | $\boldsymbol{(1/2)|7|}$, |
| | $\boldsymbol{|7(1/2)|}$ | ***and*** | $\boldsymbol{7|1/2|}$. |

***Make a conjecture.***

---

In each case the two quantities are equal. The correct conjecture is that $|ax| = a|x|$ if $a$ and $x$ are real numbers and $a \geq 0$. We will also accept this property as true.

Now we are ready to state the theorem about the relationship between the two types of sums we have defined.

THEOREM. If a series has a "sum" of $S$, then it also has a "sum according to the first arithmetic mean" and this sum is also $S$.

*Proof.* We must show that for any $1/10^K$ there is an $M$ such that the difference between $t_N$ and $S$ is less than $1/10^K$ for all $N \geq M$. Another way to say this is that for all $N \geq M$, $|t_N - S| < 1/10^K$.
At first it seems almost impossible to move from what we know, that $|s_N - S|$ becomes very small, to what we must show,

$$\left|\frac{s_1 + s_2 + \ldots + s_N}{N} - S\right| < 1/10^K.$$

Let's see if there is some way to change this expression involving the average to look more like what we know.

$$\begin{aligned}
|t_N - S| &= \left|\frac{s_1 + s_2 + \ldots + s_N}{N} - S\right| \\
&= \left|\frac{s_1 + s_2 + \ldots + s_N}{N} - \frac{N \cdot S}{N}\right| \\
&= \left|\frac{(s_1 - S) + (s_2 - S) + \ldots + (s_N - S)}{N}\right|
\end{aligned}$$

From the first property of absolute value extended to $N$ terms it follows that

$$|t_N - S| \leq \left|\frac{s_1 - S}{N}\right| + \left|\frac{s_2 - S}{N}\right| + \ldots + \left|\frac{s_N - S}{N}\right|$$

We want to show that after a certain point this right hand expression becomes and remains less than $1/10^K$ so certainly $|t_N - S|$ is less than $1/10^K$. We know that the numerators at the end of the right hand expression become very small. Let's make sure they are even smaller than the required number. We know for example that there is an $R$ such that $|s_N - S| < 1/10^{K+1}$ for all $N \geq R$. We then have

$$\begin{aligned}
|t_N - S| &= \left|\frac{s_1 + s_2 + \ldots + s_{R-1} + s_R + \ldots + s_N}{N} - S\right| \\
&\leq \left|\frac{s_1 - S}{N}\right| + \left|\frac{s_2 - S}{N}\right| + \ldots + \left|\frac{s_{R-1} - S}{N}\right| + \left|\frac{s_R - S}{N}\right| + \ldots \\
&\quad + \left|\frac{s_N - S}{N}\right|
\end{aligned}$$

$$\leq \frac{1}{N}(|s_1 - S| + |s_2 - S| + \ldots + |s_{R-1} - S|) +$$

$$\frac{1}{N}(|s_R - S| + \ldots + |s_N - S|)$$

$$< \frac{C}{N} + \frac{1}{N} \cdot \frac{(N-R+1)}{10^{K+1}}$$

where $C = |s_1 - S| + |s_2 - S| + \ldots + |s_{R-1} - S|$. Since we can always pick $R \geq 2$, the expression $(N-R+1)/N < 1$, so $(1/N)[(N-R+1)/10^{K+1}] < 1/10^{K+1}$. The numerator $C$ in the first term is some fixed positive number (depending on $R$), so if we pick $N$ large enough, say $N \geq C \cdot 10^{K+1}$, then $C/N \leq C/C \cdot 10^{K+1} = 1/10^{K+1}$. Now let $M$ be the larger of $C \cdot 10^{K+1}$ and $R$. Then for all $N \geq M$, we know that $N \geq C \cdot 10^{K+1}$ and $N \geq R$. It follows that

$$|t_N - S| < \frac{C}{N} + \frac{1}{N} \cdot \frac{(N-K+1)}{10^{K+1}}$$

$$|t_N - S| < \frac{1}{10^{K+1}} + \frac{1}{10^{K+1}}$$

$$|t_N - S| < \frac{2}{10^{K+1}}$$

$$|t_N - S| < \frac{1}{5 \cdot 10^K}$$

$$|t_N - S| < \frac{1}{10^K} \quad \blacksquare$$

## Exercises

**1.** Show that the ordinary sum, "sum," and "sum according to the first arithmetic mean" of $1 + 2 + 3 + 4$ is 10. This illustrates that in generalizing the concept of sum, the original meaning has not been lost.

**2.** Show that 0.999... converges to 1 according to the first arithmetic mean.

**3.** Show that $1 + 0 + (-1) + 1 + 0 + (-1) + \ldots$ does not converge according to the definition of "sum" in Section 3. Show that this same series does converge according to the first arithmetic mean.

***4.** Show that $1 + (-3) + 5 + (-7) + 9 + (-11) + \ldots$ does not converge according to either definition.

***5.** Define convergence according to the *second arithmetic mean* as follows: If $a_1 + a_2 + a_3 + \ldots$ is an infinite series with partial sums $s_1 = a_1$, $s_2 = a_1 + a_2$, $s_3 = a_1 + a_2 + a_3, \ldots$ and averages $t_N = (s_1 + s_2 + \ldots + s_N)/N$, then let $u_N = (t_1 + t_2 + \ldots + t_N)/N$. The series is said to converge to $U$ according to the second arithmetic mean if for any $1/10^K$ there is an $M$ such that $|u_N - U| < 1/10^K$ for all $N \geq M$. Does $1 + (-3) + 5 + (-7) + 9 + (-11) + \ldots$ seem to converge according to the second arithmetic mean? If not, why not? If it does seem to converge, what number does it converge to?

## 6. WHAT HAVE WE LEARNED FROM INFINITE SERIES?

We began this chapter by considering the sum of a finite series and then asked if this concept could be extended to an infinite series. Does $1 + 1/2 + 1/4 + \ldots$ have a "sum"? The intuitive notion was very much dependent on ideas involving infinity. We looked at the *infinitely* many partial sums and asked if they got within an *infinitely small* distance from some fixed point. For the particular series above the partial sums get closer and closer to 2, so we said it converged to 2, had a limit of 2, or had a "sum" of 2.

The amazing thing which we have been able to accomplish in the formal mathematical approach was to answer the question *without* using the notion of *infinity.* The infinite series $a_1 + a_2 + a_3 + \ldots$ with partial sums $s_1 = a_1$, $s_2 = a_1 + a_2$, $s_3 = a_1 + a_2 + a_3, \ldots$ converges to $S$ if for any $1/10^K$ there is an $M$ such that $|s_N - S| < 1/10^K$ for all $N \geq M$. Notice that the condition for convergence is stated completely in finite terms.

When calculus was first invented by Newton and Leibniz near the end of the 17th century, their work depended much on the notion of a "sum" of an infinite series. Their approach to infinite sums was closer to our intuitive approach than the formal definition—so it was much involved with infinity. The other basic notions of calculus which they developed (differentiation and integration) also involved infinity. Because of the mistakes that can easily be made when dealing with concepts involving infinity (see Chapter 6), analysis was not considered to be on a solid foundation until it had been stated in rigorous terms in the 19th century.

In this chapter we also have an example of the mathematician's constant movement toward *generalization.* Can the sum of a finite series be generalized to the "sum" of an infinite series? Since all infinite series do not have a "sum," can the notion be further generalized to include more infinite series? Can this second generalization be further generalized to include still a larger class of series? In each case the answer is affirmative only if the new generalization causes no contradictions, is a meaningful definition (does actually extend the notion to some new series), and preserves the less general notion. We have seen that this is indeed possible.

Although the mathematician usually looks for a generalization merely for the sake of its intrinsic elegance, his discoveries are often useful. The method of finding a sum by an arithmetic mean has been useful in the theory of Fourier series—one of the basic tools used in acoustics and thermodynamics.

The notion of *limit* (or convergence) is the main idea underlying calculus. It was invented (as many mathematical tools were) to help analyze a physical phenomenon—*change.* The increasing speed of a rocket after lift-off, the increase or decrease of power in an electrical system, expansion due to heat, the fluctuating rate of flow of water through a pipeline, variations in noise level are but a few examples of change.

In calculus two operations, *differentiation* and *integration,* are used to analyze change. These two operations are the *inverses* of each other, as addition and subtraction are inverses (one "undoes" what the other "does"). Both processes, differentiation and integration, are limit processes. We will not attempt to define these notions here, but the curious may want to read Sawyer's book listed below.

## Exercises

**READ FROM THE FOLLOWING**

Bergamini, David, and the Editors of Life, *Mathematics,* Life Science Library, Time Incorporated, New York, 1963, pp. 105–125.

Courant, Richard, and Herbert Robbins, *What Is Mathematics?,* Oxford University Press, New York, 1941, pp. 398–421.

Sawyer, Walter W., *What is Calculus About?,* Random House, New York, 1961.

Waismann, Friedrich, *Introduction to Mathematical Thinking,* Harper Torchbooks, Harper and Brothers, New York, 1959, pp. 123–152.

The articles on Series and Calculus, Differential and Integral, in a recent edition of the *Encyclopaedia Britannica.*

### 7. WHAT IS ANALYSIS?

The beginning of analysis is usually said to be with the work of Newton and Leibniz. Like all other great ideas, however, it was built on the work of others. The ancient Greeks had a method for finding the area of a circle which is similar to integration. Special cases of integration had been accomplished by several mathematicians before Newton and Leibniz. Differentiation, which can be interpreted geometrically as finding the tangent to a curve, had also been solved in several specific cases. By the end of the 17th century the time was ripe for someone to notice that these two operations could be generalized and were inverses of each other. It is not surprising that almost simultaneously, but independently, two great men, Newton in England and Leibniz in Germany, took the big step.

The 18th century saw many further developments including partial differentiation, and the theory of differential equations. In the 19th century the basic notions were put on a rigorous foundation and the notions of calculus were extended to the complex numbers (numbers of the form $a + bi$ where $a$ and $b$ are real numbers and $i = \sqrt{-1}$). The 20th century saw further generalizations of integration and the development of abstract analysis (functional analysis). Abstract analysis is the axiomatization and generalization of the older analysis (called classical analysis). Often it is possible to obtain several theorems of classical analysis as special cases of a single theorem in abstract analysis. The process of abstraction often clarifies things by removing extraneous details and reveals unsuspected relationships.

Although every branch of mathematics has applications to the physical world, there is no question that analysis is the part of mathematics most frequently used by scientists and engineers. Every science and engineering student takes several years of college mathematics, most of which is some form of analysis. Mechanics, optics, hydrodynamics, electricity, acoustics, and thermodynamics all use analysis. Modern cars, planes, dams, radios, television sets, refrigerators, rockets, and the rest of our technological equipment would not be if it were not for analysis.

## Exercises

**READ ONE OR MORE OF THE FOLLOWING**

Bers, Lipman, "Complex Analysis," Edited by COSRIMS, M.I.T. Press, Cambridge, Mass., 1969, pp. 7–20.

Boyer, Carl B., *A History of Mathematics,* John Wiley, New York, 1968, pp. 429–452.

*Cohen, I. Bernard, "Isaac Newton," *Scientific American,* Vol. 193, No. 6, December, 1955, pp. 73–80.

More, L. T., *Isaac Newton, a Biography,* Dover, New York, 1962.

The articles on Analysis, Sir Isaac Newton, and Gottfried Leibniz in a recent edition of the *Encyclopaedia Britannica.*

## Test

**1.** Consider the series $3 + 7 + 11 + 15 + \ldots$ with $N$ terms.

(a) What is the 5th term?
(b) What is the $N$th term?
(c) Find a formula for the sum of the first $N$ terms.
(d) What is the sum of the series if $N = 100$?

**2.** Find the first eight partial sums of each of the following infinite series and describe their behavior.

(a) $2 + 2 + 2 + \ldots$
(b) $(-5) + 5 + (-5) + 5 + \ldots$
(c) $1/2 + 1/4 + 1/8 + 1/16 + \ldots$

**3.** What is the $N$th partial sum of the series in 2(c)?

**4.** Prove that the series in 2(c) converges.

**5.** Without calculating any averages, what does $1 + 2 + 3 + 0 + 0 + \ldots$ converge to according to the first arithmetic mean? Why?

**6.** What outside reading did you do about analysis? Summarize anything additional to the text which you learned.

Don Renner—Photo Trends

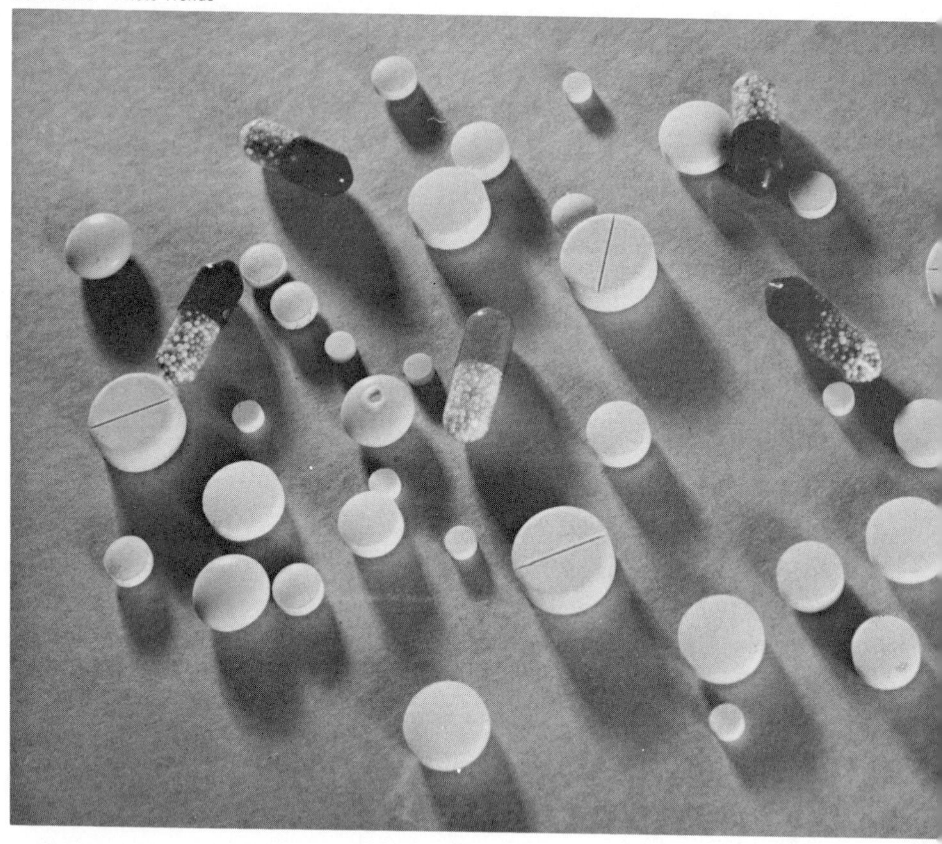

**Probability theory is given practical application in the quality control of medications and in decisions concerning possible undesirable side effects.**

8

# PROBABILITY

## Predicting The Behavior of Collections

There is one branch of mathematics which seems to touch our lives more than any other—probability and its related data-collecting science of statistics. Although probability theory affects our lives so much, it is a relatively young and somewhat unsettled part of mathematics. To keep things understandable we'll look at the original 18th and 19th century view of probability. Today this approach has been refined, but even in its original form it was extremely useful to scientists. How probability inspired a scientific theory, and how it was used to predict accurately the behavior of *collections* is the subject of this chapter.

## 1. CERTAIN BEHAVIOR AND VARIABLE BEHAVIOR

There are some events in life we consider to be certain. Let's imagine that we have a bushel of apples. If we hold up any one of the apples and release it, it will certainly fall. Or imagine fifteen men on the earth each confronted with a 25,000-lb weight. It is certain that none of them can lift the weight. We will say that *behavior is certain* if each object in a collection gives the same outcome when subjected to a fixed condition.

***In each of the two examples of certain behavior above, what is the collection, the condition, and the outcome?***

We could summarize our answer as follows:

| Example | Collection | Condition | Outcome |
|---|---|---|---|
| 1 | a bushel of apples | each apple is held in the air and released | each apple falls |
| 2 | fifteen men | each man is asked to lift a 25,000-lb weight | each man cannot lift the weight |

***Describe two more examples of certain behavior. Be sure to specify the collection, the condition, and the outcome.***

We may have some difficulty finding examples. There are two reasons. One is that some examples of certain behavior sound silly because they are so certain. If five cars all have empty gas tanks, none will start. We are so sure of this that we feel like fools even to mention it.

Other examples of certain behavior which we would like to use present another problem. We feel that these examples are certain, but cannot rule out the possibility of a different outcome. If we consider 100 people born 120 years ago, we are quite sure that none is alive today. Should we, however, hear that one of these people is alive we would be willing to believe it. This second example then is *not* one of certain behavior.

As we try to find examples of certain behavior, it becomes clear that many interesting things in life are not certain. There are, of course, many examples of behavior for which the degree of certitude is very high. If we feel it is high enough we act as if it is certain. Probability theory can help us make decisions which deal with uncertain, or variable behavior. We will say that *behavior is variable* if the objects in a collection do not all give the same outcome when subjected to the same condition.

Some examples of variable behavior are:

1. A freshman at this college this year may or may not graduate in four years.
2. A card picked from a standard deck may be a heart, a club, a spade, or a diamond.
3. If fifty people are given penicillin, some may have no allergic reaction, some may break out in hives, and some may have a severe reaction.
4. If we have several phone books and in each one we look at the sixteenth surname on the thirty-third page, it may have an odd number of letters in it and begin with a vowel, it may have an odd number of letters and begin with a consonant, it may have an even number of letters and begin with a vowel, or it may have an even number of letters and begin with a consonant.

***In each of the four examples above name the collection, the condition, and the set of outcomes being considered.***

It is possible that the same collection and condition can have either certain behavior or variable behavior, depending upon which set of outcome we observe. A coin held up and dropped several times is certain to fall each time, but it is variable how it lands—possibly the head shows, possibly the tail.

Since we can observe different sets of outcomes for the same collection and condition, we must be careful that we have a reasonable set in mind. The set we decide to observe must be such that every object in the collection *will have one* of the listed outcomes and it will have *only one* of the listed outcomes. This is called a *permissible set of outcomes* (some authors call it a sample space). We will call a member of this permissible set an *outcome* (some authors call it a simple event).

***If a card is to be picked from a standard deck, what is wrong with choosing {red, spade, even} as a set of outcomes?***

We might pick the ace of clubs from the deck and it is neither red, a spade, nor an even number. Or we might pick the four of hearts which would have two of our listed outcomes at the same time.

***Give a permissible set of outcomes to observe if we are tossing a nickel and a dime.***

A permissible set is {same, different}, if we observe the upward faces. Another set is $\{Hh, Th, Ht, Tt\}$, if $H$ stands for the head of the nickel, $T$ for the tail of the nickel, $h$ for the head of the dime, and $t$ for the tail of the dime.

## Exercises

**1.** The following proverbial expressions have literal meanings which speak of certain behavior. In each case, find a suitable collection and a condition which produces this certain behavior. Specify the certain outcome.

(a) You reap what you sow.
(b) What goes up, comes down.
(c) Birds of a feather flock together.
(d) Tomorrow is another day.
(e) Two and two are four.

**2.** The following sentences all speak of variable behavior. In each case find a suitable collection, specify the condition, and list a permissible set of outcomes.

(a) People are of different heights.
(b) A birthday can be any day of the year.
(c) Some kernels of popcorn do not pop.
(d) The sex of a family's first child is random.
(e) A falling matchbook can land various ways.
(f) Light bulbs last for various amounts of time.

**3.** For the following examples of variable behavior, several sets of outcomes are given. Which sets are permissible? If a set is not permissible, what is wrong with it? Correct it.

(a) A card is drawn from a standard deck and it is:
1. {heart, club, spade}
2. {A, 2, 3, 4, 5, 6, 7, 8, 9, 10, J, Q, K}
3. {red, spade, club}
4. {red, diamond, club}

(b) Two dice are thrown and the sum of the upward faces is:
1. {even, odd}
2. {1, 2, 3, 4, 5, 6, 7, 8}
3. {less than 6, more than 6}
4. {7 or 11, other than 7 or 11}

(c) A girl enrolled at a four-year college may be:
1. {in a sorority, a class officer}
2. {left-handed, right-handed}
3. {a freshman, a sophomore, a junior, a senior, a special student}
4. {an only child, an only daughter, a first child}

**4.** Each of the following is an example of a collection and a condition which has certain behavior or variable behavior, depending upon what is being observed. Name a certain outcome for each. For each example find a permissible set of outcomes for some variable behavior.

(a) Ten people go swimming.
(b) One hundred people each smoke a pack of cigarettes a day for ten years.
(c) A pointer mounted on a clocklike card marked 1,2,3,4,5,6 is spun one hundred times.
(d) Fifty listed people are alive today.

## 2. THE LAPLACE DEFINITION OF PROBABILITY

At this point we should understand the meaning of certain behavior and variable behavior, and know how to decide if a set of outcomes is permissible. Now let's consider a question that was asked as early as the 18th century.

By the 18th century many types of certain behavior had been studied by physical scientists. Their technique was to find mathematical relationships that matched phenomena being observed and to use these mathematical relationships to predict new information about the physical situation. The falling of a released apple is one such example. Scientists found that this certain event had some numerical properties. The velocity $v$ of the falling apple $t$ seconds after release was equal to approximately $32t$ feet per second. The formula $v = 32t$ is a mathematical relation which matches this property. This formula can be used to predict the velocity of the apple (or any other object) at a specific time after release. It can also be used in connection with other known formulas to predict even more complicated behavior. For example, if we throw an apple straight up with initial speed of 16 feet per second, it is possible to predict accurately how high it will go.

During the 18th century mathematicians and scientists (most mathematicians were both) began to wonder if there were some mathematical relationships which matched variable behavior. If so, perhaps mathematics could be used to predict variable behavior.

***Perform the following experiment:***
***(1) Toss a penny ten times, keep track of the number of heads and divide this by the number of tosses.***
***(2) Toss the penny ten more times, keep track of the number of heads, add this to the number of heads in the previous step, and divide the total number of heads by the total number of tosses.***
***(3)–(5) Repeat Step (2).***

Our tabulation might look as follows:

| Step | Heads in 10 Tosses | Total Heads | Total Tosses | Heads/Tosses |
|---|---|---|---|---|
| 1 | 3 | 3 | 10 | $3/10 = 0.30$ |
| 2 | 6 | 9 | 20 | $9/20 = 0.45$ |
| 3 | 7 | 16 | 30 | $16/30 = 0.53$ |
| 4 | 2 | 18 | 40 | $18/40 = 0.45$ |
| 5 | 6 | 24 | 50 | $24/50 = 0.48$ |

***As the number of tosses increases, how does the number of heads divided by the number of tosses seem to vary?***

This short experiment seems to indicate that the number of heads divided by the number of times the coin is tossed fluctuates about 1/2, or 0.50 in decimals. Our feeling is that if we would repeat the experiment more times the ratio would get closer to 1/2.

***If you are working with other people, combine your final results in the last table to find the number of heads divided by the number of tosses for 100, 150, 200, 250, and 300 tosses. Does our conjecture seem to be true?***

Penny tossing is an example of a collection and a condition which leads to variable behavior. The collection is our set of tosses, and the fixed condition is that a fair coin is tossed the same way. The behavior is variable because the outcomes are different for the individuals of the collection. When we view the collection there does indeed seem to be some rule in the variableness. For a large collection of tosses (although there is no definite way to say how large) experience verifies that heads land up approximately half of the time. Admitting that there is some vagueness in this notion of "large collection" we will say that for a large collection of tosses, coins land heads up 1/2 of the time—or the *probability* of the head landing up is 1/2. Notice that up to this point in this chapter we have not been doing mathematics as such; we have been dealing with the physical world.

The mathematical theory of probability was inspired by experiences like this one. After observing heads roughly half of the time, the mathematicians asked, "Is there some mathematical relation which mirrors the fact that we get heads half of the time? Could we by-pass the experiment and predict the behavior from some matching mathematics?"

***Can you see any mathematical relationship which matches the fact that heads land up approximately half of the time?***

Considering the permissible set of outcomes, either the head or the tail appears and, provided the coin is honest, each case seems *equally likely.* Head, the favorable outcome, is one of the two possibilities. The number of favorable outcomes divided by the number of possible outcomes in this example is 1/2, the same as the experimental probability. It seems that the formula $P = F/T$, *probability equals the number of favorable outcomes divided by the total number of outcomes,* mirrors the actual behavior of a large collection. Inspired by this formula mathematicians began to study probability in an abstract way. To a mathematician, the permissible set of outcomes with equal likelihood is a set of symbols and the probability of any subset of that permissible set is the number of elements in the subset divided by the number of elements in the permissible set. The job of deciding if a set is permissible, if the outcomes are equally likely, and which outcomes form the favorable subset belongs to the scientist.

***Apply the definition of abstract probability to a nickel and a dime to find the probability of two heads landing up if a nickel and a dime are tossed.***

We let $H$ represent the head of the nickel, $T$ the tail of the nickel, and use $h$ and $t$ in a similar way for the dime; then we assume that the combinations $Hh$, $Ht$, $Th$, and $Tt$ are all equally likely to occur. On the abstract level, the set we consider is $\{Hh, Ht, Th, Tt\}$. The only favorable case is $Hh$, so abstractly we want the probability of $\{Hh\}$ which is 1/4. Therefore, we apply this to the real world and say that the probability of tossing two heads is 1/4. In practice, during the early stages of probability, people did not bother with explicitly moving to the abstract level and back again to the concrete situation. They simply considered the concrete situation and used $P = F/T$. At this point in our discussion we will also use this simple approach, remembering, however, that what we are actually involved in is not just mathematics, but mathematics applied to a physical situation.

***What is the probability of tossing one head and one tail if a nickel and a dime are tossed?***

The probability of tossing one head and one tail is $2/4 = 1/2$ because there are two favorable outcomes $Ht$ and $Th$ out of a total of four possible outcomes.

This approach to probability developed during the 18th and 19th centuries and was unified and advanced most by the work of Laplace. The formula $P = F/T$ is often called the Laplace definition of probability.

As far as physical problems are concerned, there are some drawbacks to this definition. To say that the outcomes are "equally likely" or "equally probable" and then use this to define probability seems to be circular. Besides this, it is also an idealization of the real world. It assumes that we can always reduce a problem to a permissible set of outcomes which are "equally likely." There are certainly situations where the possibility of some outcomes outweigh the others and a reduction to "equally likely" events cannot be made.

One example in which the outcomes are not "equally likely" would be a coin which has been weighted, deliberately constructed to favor one side. It still makes sense to speak of the probability of the head landing up, but in the case of this false coin it would be better to assign the probability by experimentation. In actual experiments the number of favorable cases divided by the total number of cases always begins to cluster around some fixed number (as in our penny tossing experiment). This fixed number is then called the probability of the outcome being considered. The assumption that the ratios in an experiment will always begin to cluster around some fixed number is called *The Law of Large Numbers.*

When there is no reason to believe that the outcomes are not equally likely, we work with Laplace's idealized probability. In these cases, the predictions usually turn out to be amazingly accurate. This should not surprise us. In geometry we idealize the imperfect circles we find around us. We reason about perfect circles, and then apply the results to the real world. These results are very useful, even though in reality every "round" object is slightly imperfect.

Another difficulty that arises from the Laplace approach is in the layman's interpretation of this definition of probability. It tends to make him concentrate on the outcomes alone and to forget about the collection. When the weatherman states that the probability of precipitation today is 20% and our picnic is rained out, we are furious at what we call inaccurate forecasting. The truth is that the weatherman never did say that it would not rain today. What he did say is that given a large collection of days, each with the same atmospheric conditions as today, then 1/5 of the time it will rain. Whether it rains today or does not rain, the weatherman is right!

We can best sum up the difficulties of the Laplace definition by saying that it is an oversimplification. It works well with the very simple situations of some games of dice and cards. The real world is usually quite complex and a more sophisticated approach is often needed. In spite of its simplicity, the Laplace approach has played a large part in the development of science. In Section 3 we will see how it inspired Mendel's First Law.

## Exercises

**1.** Toss two pennies twenty-four times. The possible outcomes are two heads, one head and one tail, and two tails (if we do not distinguish between the two pennies). Keep track of the number of times each outcome occurs. To obtain an experimental probability for each outcome, divide the number of occurrences of each outcome by 24 and change this fraction to the closest two-placed decimal.

2. If two pennies are tossed, use the Laplace definition to predict the probability of throwing each of two heads, one head and one tail, and two tails. (To consider equally likely outcomes you must use the set of outcomes $\{HH, HT, TH, TT\}$.) Were your results in Exercise 1 approximately the same?

3. An ordinary six-faced die is thrown. What is the probability of:
   (a) The five face landing upward?
   (b) Either 1 or 3 landing upward (that is, both 1 and 3 are favorable cases)?
   (c) An even number landing upward?
   (d) A number less than 6 landing upward?
   (e) One of 1, 2, 3, 4, 5, or 6 landing upward?

4. A card is picked from a standard deck. What is the probability of drawing:
   (a) An ace?
   (b) A heart?
   (c) A picture card?
   (d) A card with number less than 5? (Count the ace as less than 5.)

5. In 1960 in the United States, 26 children out of every 1000 born alive died within their first year of life. Assume that in another year the conditions are similar. Calculate the probability that a child born in that year will live longer than a year.

6. According to one insurance mortality table, the probability of a person aged 20 dying within the next year is 0.0018. What does this mean? If John Smith is aged 20, can he draw any conclusion from this table about whether he will die in the next year?

7. The probability that a newborn child is a boy is 0.51. If the Smiths have already had two girls, what is the probability that their next child will be a boy?

## 3. PROBABILITY AND MENDEL'S FIRST LAW

About the middle of the 19th century, Gregor Mendel noticed a strange thing which happens when plants are artificially fertilized. It is possible to cross-pollinate two pea plants, one tall and the other a dwarf plant, and get hybrid plants, all of which are tall. This is contrary to intuition; intuition says that the descendants should have medium height. Instead of a blending, the hybrid generation resembles one of the parents completely. However, if these hybrids are then self-pollinated, the next generation will be composed of some tall plants and some dwarf plants. Before Mendel, no one had undertaken the huge task of raising enough plants under controlled conditions to determine whether there was some law governing the offspring of hybrids. Mendel's experiment took seven years.

Mendel, besides being a botanist, was aware of progress in other fields. He knew that the physical scientist had been able to use mathematics to predict physical behavior and hoped to do a similar thing in natural science. He was also aware of the mathematics being developed and could have had a feeling from the start that the theory of probability was the mathematics which would match the behavior of heredity.

Mendel chose for his experiment several varieties of pea plants because they were easy to grow, matured quickly, were easy to pollinate artificially, easy to protect from foreign pollen, and had fertile hybrids. The pea plants also had several characteristics that differed in some of the varieties: size, arrangement of flowers, color of seed (endosperm), color of seed coat, shape of seed, shape of pod, and color of the unripe pod. The pea plants would then provide material for several different experiments. Let's look at just one of his series of experiments—the one in which he concentrated on the color of the seed.

Some pea plants have seeds with a yellow endosperm, others a green endosperm. Pure-line seeds can be obtained from plants that, when self-pollinated, breed true. Mendel planted equal numbers of pure-line yellow and pure-line green seeds. He cross-pollinated these pure-line plants in pairs of one yellow and one green. He used the pollen from each member of the pair on the other. In all cases the plants produced yellow seeds. These were hybrid seeds, but resembled only one parental plant. Mendel said that yellow was dominant over green for these seeds. Since both types of plants produced yellow seeds, he concluded that whatever factors were at work, it was immaterial whether they came from the pollen-bearing or the seed-bearing plant.

The main part of his experiment began when he planted the hybrid seeds. From this point on he would do no more cross-pollination. Each plant would be self-pollinated artificially, and protected from all foreign pollen. He raised 258 plants from the hybrid yellow seeds—his parental generation. (See Fig. 8–1.)

The parental hybrid plants produced in the first filial generation a mixture of 6022 yellow seeds and 2001 green seeds. The ratio of the dominant color to the recessive was about 3 to 1. (When Mendel worked with the other differing characteristics, such as size or flower arrangement, a similar ratio occurred.)

Whether the first filial generation seeds were hybrid or pure-line could only be determined by the second filial generation. All green seeds produced plants which yielded only green seeds. All green seeds were pure-line. Mendel raised 519 plants from yellow first-filial-generation seeds. Of these plants 166 produced all yellow seeds and 353 produced both yellow and green seeds. This meant that among the first-generation yellow seeds, the ratio of pure-line to hybrid was about 1 to 2. Looking at

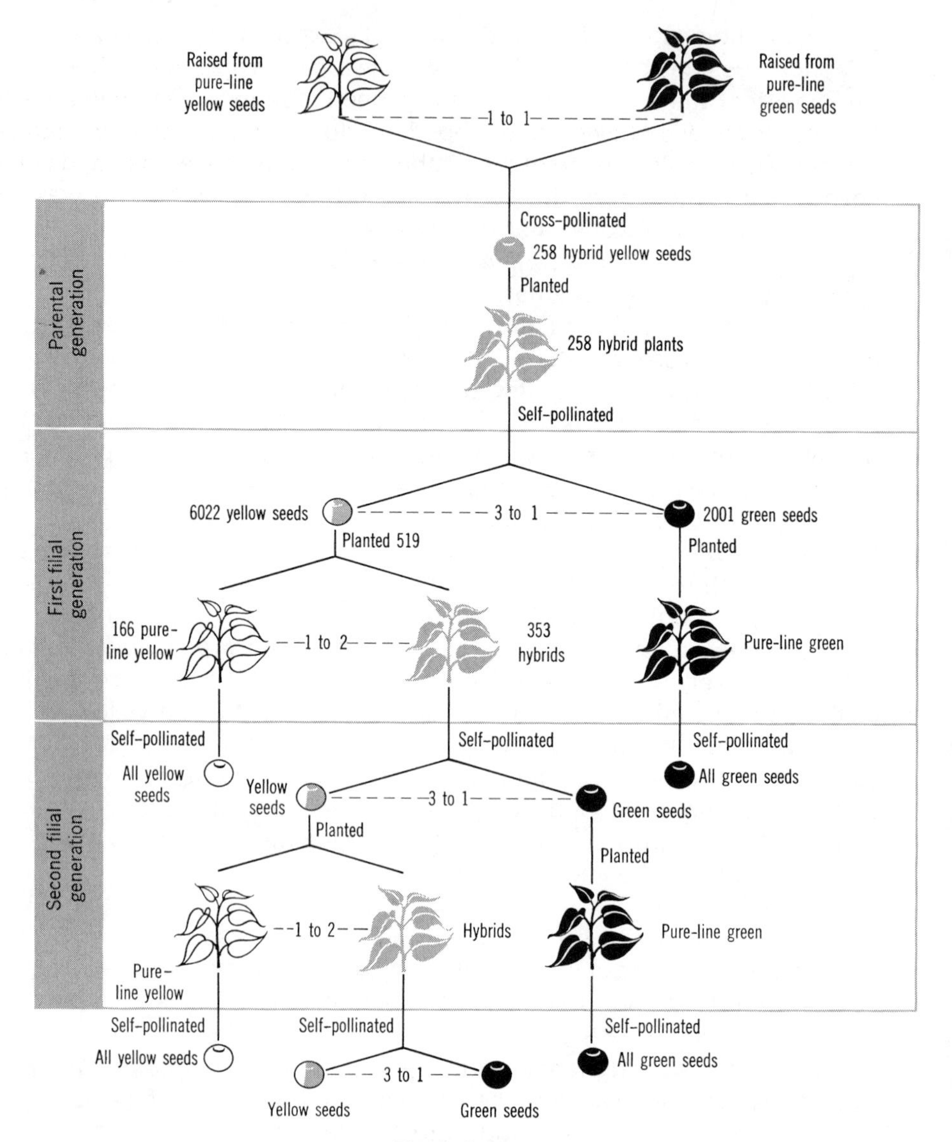


**Fig. 8–1.**

*all* of the first-generation seeds, Mendel saw that about 1/4 of them were pure-line yellow, about 1/2 were hybrid (and were yellow), and about 1/4 were pure-line green.

Mendel then found that this process repeated itself in the second filial generation raised from the seeds of the first-generation hybrid plants. In fact, he followed six generations and found that seeds from hybrid plants consistently give about 1/4 pure yellow seeds, about 1/2 hybrid seeds, and about 1/4 pure green seeds.

What Mendel wanted to do then was to formulate a scientific theory of what was happening—an explanation and some matching mathematics which could be used to predict the future hereditary properties of these plants and, hopefully, even other organic forms.

***Where have you seen this division into 1/4, 1/2, 1/4 before?***

If two coins are tossed many times, we know that two heads will appear approximately 1/4 of the time, one head and one tail about 1/2 of the time, and two tails appear about 1/4 of the time. We get these numbers both from experimentation and from the Laplace definition of probability. Mendel certainly knew this in 1856 when he began his pea plant experiment. We don't know exactly how he came to his final conclusion, but it seems likely that he could have noticed that plant behavior resembled the tossing of coins (see Fig. 8–2).

The new generation from hybrids resembles the two upward faces of the tossed coins—so there must be two factors (now called genes) at work, one received from each gamete (pollen or egg cell). The upward face of the coin, in turn, is one of two possible faces, so the gamete must be passing on one of two factors which occur in the parental plant. (In the illustration $Y$ stands for the yellow factor, $G$ for the green.) Mendel assumed that it was "equally likely" which factor the gamete received and

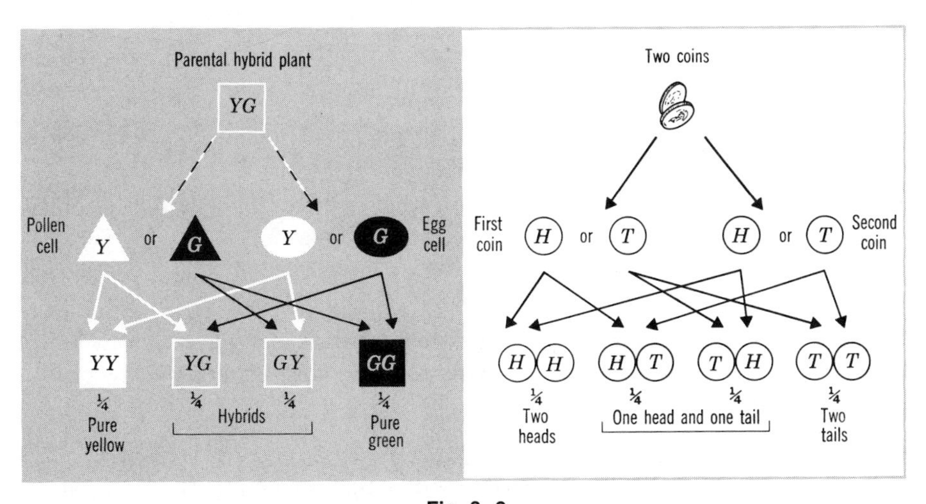


**Fig. 8–2.**

"purely a matter of chance which types of cells unite." He said, "According to the laws of probability the various combinations will on the average occur equally often."

This explanation of the transmission of characteristics is called Mendel's First Law. Because Mendel published his work in a relatively obscure journal, it was generally overlooked during his lifetime, but it was rediscovered at the beginning of the 20th century. Today there are some refinements of Mendel's theory, but it is still accepted as the basic process underlying heredity in every form of life from virus to human. Mendel's discovery was certainly one of the great events in the history of biology.

For the mathematician, Mendel's work was also a big step. At that time, probability was not a completely respectable branch of mathematics. It seemed to be a curiosity built on the shaky ground of concepts like "for a large number" and "equally likely." It was sometimes useful to gamblers, had been used to some degree by insurance companies to predict life expectancy, and was used to weigh legal evidence. It had *not* been used by any of the physical or natural sciences. It is amazing that Mendel had the insight to see that this new branch of mathematics was exactly what was needed to mirror genetics. In the fifty years following Mendel's experiment, probability would be used in the study of heat, gases, and radiation. By the early 20th century both probability and Mendel's First Law would be respected and the time would be ripe to apply probability to make some accurate predictions about future generations.

In Section 4 we will develop some further properties of probability and in Section 5 we will use these to make the predictions.

## Exercises

**1.** Using Mendel's view of genes, explain why his cross-pollination of pure lines led to all hybrids.

**2.** (a) In a hybrid yellow-green seed plant, what is the probability that a gamete will carry the gene for yellow? The gene for green?

(b) In a pure-line yellow seed plant, what is the probability that a gamete will carry the gene for yellow?

**3.** Brown eyes are dominant over blue eyes. Assume that when a brown-eyed person marries a blue-eyed person their children have either definitely blue or definitely brown eyes.*

*There are cases of people who are genetically pure brown-eyed, but in fact have blue eyes. This phenomenon is called incomplete penetrance. Such people will pass a gene for brown eyes to their children. You may ignore these unusual cases when answering Exercise 3.

(a) If a genetically pure brown-eyed person marries a genetically pure blue-eyed person, what is the probability that if they have a child he will be brown-eyed?

(b) If the couple in part (a) have a brown-eyed child and he marries a genetically pure blue-eyed person, what is their probability of having a brown-eyed child?

(c) Assume that the couple in part (a) have a brown-eyed child who marries a brown-eyed person who carries a recessive gene for blue. What is their probability of having a blue-eyed child?

**4.** If Mendel had allowed his hybrid parental plants to cross-pollinate at random, would he have had a different result than he obtained by artificial self-pollination?

**5.** If Mendel had begun with equal numbers of pure-line yellow and pure-line green pea plants and had allowed cross-pollination to occur at random (assume that conditions are such that a truly random cross-pollination does occur), what results would he have obtained?

**6.** In pea plants, yellow seeds are dominant over green seeds and smooth seeds are dominant over wrinkled. Mendel found that if two characteristics are watched in several generations of pea plants, they each follow his First Law independent of the other. If hybrid seeds are obtained by crossing pure-line yellow smooth plants with pure-line green wrinkled plants, how will the hybrid seeds look? If plants are raised from these hybrid seeds and self-pollinated, what will be the ratios of yellow smooth seeds to yellow wrinkled seeds to green smooth seeds to green wrinkled seeds? (Hint: Use *YYSS* and *GGWW* to symbolize the original plants. Symbolize the hybrids a similar way. List all of their possible gametes, and then all possible offspring.)

**7.** Read from the following:

Mendel, Gregor, "Mathematics of Heredity," in *The World of Mathematics,* Vol. 2, J. R. Newman, Simon and Schuster, New York, 1956, pp. 937–949.

The article on Gregor Mendel, or the section on Transmission Genetics in the Genetics article in a recent edition of the *Encyclopaedia Britannica.*

## 4. SOME PROPERTIES OF PROBABILITY

In this section we will examine some of the properties of probability. We will think of probability as the number assigned to an outcome by the Laplace definition or the number assigned by the Law of Large Numbers. Let's begin by considering a collection, a condition, and several sets of outcomes.

The collection is a large number of spins of two spinners. One face is divided into three equal sections and colored red, white, and blue. The other face is divided into four equal parts and these are labelled 1, 1, 2, and 3. The condition is that these two spinners are spun each time in the same manner. We will assume that they are constructed so that they cannot stop on the dividing lines between the sections. There are several sets of permissible outcomes. Some of them are:

$S_1 = \{\text{spinners stop}\}$
$S_2 = \{R, W, B\}$
$S_3 = \{1, 2, 3\}$
$S_4 = \{R1, R2, R3, W1, W2, W3, B1, B2, B3\}$
$S_5 = \{\text{even, odd}\}$
$S_6 = \{R \text{ or } W \text{ and 1 or 2}, R \text{ or } W \text{ and 3}, B\}$

We are using $R$, $W$ and $B$ to stand for red, white, and blue, respectively, in $S_2$ $S_4$, and $S_6$. Even and odd in $S_5$ refer to the numbers on the second spinner.

***Use the Laplace definition of probability to complete the following table. (In the table* P(A) *means "the probability of outcome* A.")**

| Permissible Set of Outcomes | Probability of Each Outcome | | |
|---|---|---|---|
| $S_1$ | $P(\text{spinners stop}) =$ | | |
| $S_2$ | $P(R) =$ | | |
| | $P(W) =$ | | |
| | $P(B) =$ | | |
| $S_3$ | $P(1) =$ | | |
| | $P(2) =$ | | |
| | $P(3) =$ | | |
| $S_4$ | $P(R1) =$ | $P(W1) =$ | $P(B1) =$ |
| | $P(R2) =$ | $P(W2) =$ | $P(B2) =$ |
| | $P(R3) =$ | $P(W3) =$ | $P(B3) =$ |
| $S_5$ | $P(\text{even}) =$ | | |
| | $P(\text{odd}) =$ | | |
| $S_6$ | $P(R \text{ or } W \text{ and 1 or 2}) =$ | | |
| | $P(R \text{ or } W \text{ and } 3) =$ | | |
| | $P(B) =$ | | |

Since "spinners stop" is a certain outcome, we may hesitate to use the Laplace definition because we discussed it only in terms of variable behavior. However, certain behavior fits nicely into this approach. We can use $S_1$ since each outcome is equally likely (there is only one), then $P(\text{spinners stop}) = 1/1 = 1$ since there is one favorable case out of one possible case. If we prefer to think of what happens experimentally, every trial leads to the spinners stopping, so no matter how many trials are involved, the number of times the spinners stop divided by the number of trials is $N/N = 1$. It seems quite reasonable to say that if an outcome $A$ is certain, then $P(A) = 1$.

We should complete the rest of the table as follows. Using $S_2$ (the outcomes are equally likely), $P(R) = P(W) = P(B) = 1/3$. We cannot use $S_3$ for the next set since the Laplace definition requires equally likely outcomes. Instead we use $\{1_A, 1_B, 2, 3\}$ where $1_A$ represents one of the quarter sections marked 1 and $1_B$ represents the other; then $P(1) = 2/4 = 1/2$ and $P(2) = P(3) = 1/4$. In order to find the probabilities for $S_4$ we must use a new set, the equally likely set $S_7 = \{R1_A, R1_B, R2, R3, W1_A, W1_B, W2, W3, B1_A, B1_B, B2, B3\}$; then in $S_4$, $P(R1) = P(W1) = P(B1) = 2/12 = 1/6$ and all of the other probabilities are $1/12$. For $S_5$ we must use $\{1_A, 1_B, 2, 3\}$ again, then $P(\text{even}) = 1/4$ and $P(\text{odd}) = 3/4$. In order to find the probabilities in $S_6$ we must also use $S_7$. $P(R \text{ or } W \text{ and } 1 \text{ or } 2) = 6/12 = 1/2$, $P(R \text{ or } W \text{ and } 3) = 2/12 = 1/6$, and $P(B) = 4/12 = 1/3$. In our table, for each outcome $A$, $P(A)$ is a real number and $0 < P(A) \leq 1$, that is $P(A)$ is greater than 0 and less than or equal to 1.

We might wonder if there is anything which should be assigned probability zero. The most reasonable thing in the examples we have considered would be an impossible outcome. For example, what is $P(\text{spinners do not stop})$? The Laplace definition as we discussed it does not literally apply since "spinners do not stop" is not a member of any permissible set of outcomes. If we think of an experiment and spin the two spinners $N$ times, how many times will they fail to stop? We know that they fail to stop zero times, so the number of times they do not stop divided by the number of trials is $0/N = 0$. No matter what $N$ is, this number will not change. We will say that $P(\text{spinners do not stop}) = 0$. In general, if $A$ is an impossible outcome (an outcome certain not to happen), then $P(A) = 0$. We can think of this as an addition to the Laplace definition.

We should sum up what we have noticed so far as follows:

PROPERTY 1. The probability $P(A)$ of an outcome $A$ is a number and $0 \leq P(A) \leq 1$. If $A$ is impossible $P(A) = 0$. If $A$ is certain, $P(A) = 1$.

If we were studying abstract mathematical probability, we would assume this property.

***For each set of outcomes in the last table, find the sum of the probabilities of each outcome.***

We notice that the sum of the probabilities of the outcomes in each permissible set of outcomes is 1. We will call this Property 2. It would be another assumption in an abstract approach.

PROPERTY 2. If $\{A_1, A_2, \ldots, A_n\}$ is a permissible set of outcomes, then $P(A_1) + P(A_2) + \ldots + P(A_n) = 1$.

We might have noticed that the set of outcomes $S_5 = \{\text{even, odd}\}$ is related to $S_3 = \{1, 2, 3\}$. "Even" means 2 in this case, and "odd" means 1 or 3. If we are looking for $P(\text{odd}) = P(1 \text{ or } 3)$, it seems that we should be able to calculate it from $P(1)$ and $P(3)$.

***How are* P(*1*), P(*3*), *and* P(*1 or 3*) *related?***

We should notice that $P(1) = 1/2$, $P(3) = 1/4$, and $P(1 \text{ or } 3) = 3/4$, so $P(1 \text{ or } 3) = P(1) + P(3)$. Property 3 (another assumption in mathematical probability) is a general statement of this relationship.

PROPERTY 3. If $\{A_1, A_2, \ldots, A_n\}$ is a permissible set of outcomes, then $P(A_i \text{ or } A_j) = P(A_i) + P(A_j)$ if $i \neq j$. A similar relationship is true for any number of outcomes connected by "or."

***Find* P(*2 or 3*) *using Property 3.***

We should get that $P(2 \text{ or } 3) = P(2) + P(3) = 1/4 + 1/4 = 1/2$.

***Find* P(*1 or 2 or 3*) *using Property 3.***

From Property 3, $P(1 \text{ or } 2 \text{ or } 3) = P(1) + P(2) + P(3) = 1/2 + 1/4 + 1/4 = 1$. We could have approached this another way by using Property 1. Since one of the outcomes 1, 2, or 3 must occur, then this "or" expression is a certain outcome and the probability is 1.

A word of warning is in order about Property 3. As stated, it applies only to "or" expressions which have been formed from the outcomes of *one* permissible set. If we are asked to find $P(R \text{ or } 1)$ and apply Property 3 to a mixture of $S_2$ and $S_3$ we would get $P(R) + P(1) = 1/3 + 1/2 = 5/6$. This is *not* the correct probability of $P(R \text{ or } 1)$. To use Property 3 we must choose one (equally likely) permissible set of outcomes. In this problem $S_4$ will work. Since $R$ means $R1$ or $R2$ or $R3$, and 1 means $R1$ or $W1$ or $B1$, we have

$$\begin{aligned} P(R \text{ or } 1) &= P(R1 \text{ or } R2 \text{ or } R3 \text{ or } R1 \text{ or } W1 \text{ or } B1) \\ &= P(R1 \text{ or } R2 \text{ or } R3 \text{ or } W1 \text{ or } B1) \end{aligned}$$

because $R1$ or $R1$ simply means $R1$. Therefore,

$$\begin{aligned} P(R \text{ or } 1) &= P(R1) + P(R2) + P(R3) + P(W1) + P(B1) \\ &= 1/6 + 1/12 + 1/12 + 1/6 + 1/6 \\ &= 8/12 \\ &= 2/3. \end{aligned}$$

We now have a property which covers "or" expressions and it is natural to ask if there is one for "and" expressions also. When we speak of $P(W2)$, we really mean $P(W \text{ and } 2)$.

---

***How are* P(W *and* 2), P(W), *and* P(2) *related?***

---

Since $P(W \text{ and } 2) = 1/12$, $P(W) = 1/3$, and $P(2) = 1/4$, we have $P(W \text{ and } 2) = P(W) \cdot P(2)$.

---

***Label the first spinner I, II, III in the red, white, and blue sections, respectively. Let* A = *the numbers on the two spinners match. Let* C = *the number on at least one of the spinners is odd. Find* P(A *and* C), P(A), *and* P(C). *Does* P(A *and* C) = P(A) · P(C)?**

---

Using $S_7$ and thinking of $R$ as I, $W$ as II, and $B$ as III, the only favorable cases for $A$ and $B$ are $R1_A$, $R1_B$, and $B3$, so $P(A \text{ and } C) = 3/12 = 1/4$. Using $S_7$ again, $R1_A$, $R1_B$, $W2$, and $B3$ are favorable for $A$, so $P(A) = 4/12 = 1/3$. For $C$, $R1_A$, $R1_B$, $R2$, $R3$, $W1_A$, $W1_B$, $W3$, $B1_A$, $B1_B$, $B2$, $B3$ are all favorable, so $P(C) = 11/12$. Therefore, $P(A \text{ and } C) = 1/4 \neq (1/3)(11/12) = P(A) \cdot P(C)$.

***What is the difference between these last two examples?***

In the first example "$W$ and 2" is an expression involving two independent events. If the second spinner stops at 2, or if it doesn't, it has no influence on whether or not the first stops on $W$. In the second example "the numbers match" and "at least one number is odd" are not independent. If at least one number is odd (or is not odd), it influences the number of ways in which it is possible to have matching numbers. The property we want to notice concerns only *independent outcomes* (that is, outcomes from unrelated sets).

PROPERTY 4. If $A$ and $B$ are two outcomes which are independent of each other, then $P(A \text{ and } B) = P(A) \cdot P(B)$.

In practice it is sometimes difficult to decide whether or not we are dealing with independent events. One way that is used to decide this is to define two outcomes $A$ and $B$ to be independent if $P(A \text{ and } B) = P(A) \cdot P(B)$. Of course, if this is how we are deciding whether or not two outcomes are independent, then we cannot use Property 4 to calculate the probabilities. This definition of independent events is usually used in an abstract approach rather than assuming Property 4.

In Section 5 we will use these properties to make some interesting predictions about heredity.

## Exercises

**1.** Let one spinner have its face divided in half and labelled I, II. Let a second spinner have its face divided into thirds and labelled 1, 2, 3. If the spinners are spun, find:

(a) $P(\text{I})$
(b) $P(\text{II})$
(c) $P(1)$
(d) $P(2)$
(e) $P(3)$
(f) $P(1 \text{ or } 3)$
(g) $P(\text{I}2)$
(h) $P(1) + P(2) + P(3)$
(i) $P(\text{numbers match})$
(j) $P(\text{numbers do not match})$
(k) $P(\text{at least one number is even})$
(l) $P(\text{numbers match and at least one is even})$

**2.** Let three coins be tossed. Find:

(a) $P(\text{three heads})$
(b) $P(\text{two heads})$
(d) $P(\text{coins match})$
(e) $P(\text{coins do not match})$

(c) $P$(at least one tail)

(f) How could Property 2 and the answer to (d) be used to answer (e)?

**3.** Let a nickel with $P(H) = 0.80$ and a dime with $P(t) = 0.65$ be tossed. Find:

(a) $P(T)$ if $T$ stands for the tail of the nickel
(b) $P(h)$ if $h$ stands for the head of the dime
(c) $P(Hh)$
(d) $P(Ht$ or $Th)$
(e) $P$(faces match)
(f) $P(Ht$ or $Hh$ or $Tt$ or $Th)$

**4.** Let a dime with $P(h) = 2/3$ be tossed five times. Find:

(a) $P(htthh)$ if reading from left to right indicates which face lands upward for the first, second, . . . , fifth toss.
(b) $P(hhhhh)$
(c) $P(ttttt)$
(d) $P(htttt)$
(e) $P(thttt)$
(f) $P(tthtt)$
(g) $P$(only one head in five tosses)

**5.** Let *YGSW* represent a hybrid pea plant obtained by crossing a genetically pure yellow smooth seed pea plant with a genetically pure green wrinkled seed pea plant. Mendel showed that it is equally likely for this plant to have gametes *YS, YW, GS, GW.* If this plant is self-pollinated, use Property 4 to find the probability that a seed will be of the form *YGWW, YYWW, GYWW.* Use Property 3 to find the probability that a seed will look yellow and wrinkled. (Yellow is dominant, and smooth is dominant.)

## 5. PREDICTING THE HEREDITY OF FUTURE GENERATIONS

At the beginning of the 20th century, Mendel's view of the process of heredity was rediscovered and accepted. In 1908 a mathematician, G. H. Hardy, applied the principles of probability to Mendel's First Law and came up with a prediction about future generations. This was one of the earliest attempts to study *population genetics.*

Hardy's work can be applied to a characteristic that has several variations in most any form of plant or animal reproduction. We will first look at the specific example of Section 3 and follow a characteristic with two variations—the color of the seeds of a pea plant. We will then develop a general statement about a single characteristic with two variations.

Let's imagine that we have a garden of 750 pea plants which will be the parental generation; 10% of the plants are pure-line yellow seed plants, 40% are pure-line green seed plants, and the remaining 50% are hybrid yellow-green seed plants. We will represent the two factors (genes) in the pure-line yellow plant by *YY,* and in the green plant by *GG.* The left letter indicates the factor received from the seed-bearing (maternal) plant, and the right letter indicates the factor received from the pollen-bearing (paternal) plant. The hybrid plants will have factors *YG* and *GY* and we will assume that there are equal numbers of each type—25% of the hybrid plants are *YG*, and 25% are *GY*. Since these two types of hybrids act alike, we will always assume that we have equal numbers of each hybrid type.

The percentage distribution which we have just given could be thought of as a probability.

---

***If you choose a pea plant at random from the above garden, what is the probability that it will be* YY? GG? YG? GY?**

---

Since 10% of our plants are *YY*, this means that 10 out of every 100 plants are *YY*. If we think of probability as $P = F/T$, then the probability of choosing a *YY* plant is $P = 75/750 = 0.10$. Similarly, the probability of choosing a *GG* plant is 0.40, of picking a *YG* plant is 0.25, and of choosing a *GY* plant is 0.25. Because of this we will use these decimals like probabilities, and call them the distribution of types. We'll write $P_0(YY) = 0.10$, $P_0(GG) = 0.40$, $P_0(YG) = 0.25$, and $P_0(GY) = 0.25$ when we speak of the distribution of types for these parental plants.

---

***Find* $P_0(YY) + P_0(YG) + P_0(GY) + P_0(GG)$. *What property of probability does it illustrate?***

---

We are going to allow the plants to cross-pollinate at random. We will assume that conditions are such that random pollination occurs. We would like to predict the distribution of types in the first filial generation.

There will be two steps needed to solve this. First, we need to know the distribution of types for the gametes produced by the parental generation. A gamete (egg or pollen cell) could contain the *Y* factor or the *G* factor. What is $P_0(Y)$, the probability that a gamete carries the *Y* factor? What is $P_0(G)$? Second, once we have this gametic distribution, we can use it to calculate the distribution of the first filial generation, $P_1(YY)$, $P_1(YG)$, $P_1(GY)$, $P_1(GG)$. In random cross-pollination we will assume that which two gametes unite is a matter of chance.

Let's begin with the problem of finding the gametic distribution. In each plant there are two genes which determine the color of its seeds. One of the genes was passed to the plant by an egg cell; it is called the *maternal gene*. The other gene was passed to the plant by a pollen cell; it is called the *paternal gene.* When a plant produces gametes of its own they receive one of these genes. (Notice that an egg cell can receive either the maternal or paternal gene of the plant, and similarly for a pollen cell.) Mendel assumed that it was equally likely for a gamete to receive the maternal or paternal gene—that is, the probability of receiving the maternal gene is $L = 1/2$, and that of receiving the paternal gene is $R = 1/2$. We are using $L$ and $R$ to remind ourselves that we write the maternal gene on the left and the paternal gene on the right.

---

***Calculate the probability of an egg cell receiving the* Y *gene from a* YG *plant.***

---

This is an example of an "and" expression—the gamete must come from the *YG* plant *and* receive the maternal gene. Since these outcomes are independent, we can use Property 4. The probability can be obtained by multiplying the probabilities of the individual events, $P_0(YG)L = (0.25)(1/2)$.

---

***Find* $P_0(Y)$*, the probability that a gamete produced by the parental generation will carry the* Y *factor.***

---

We have already found the probability of the gamete receiving the *Y* factor from a *YG* plant (we calculated for an egg cell, but the probability is the same for a pollen cell). Now we must consider the other possibilities. A *Y* gamete could arise from a *YY* plant and receive the maternal gene, or from a *YY* plant and get the paternal gene, or it could arise from a *YG* plant and get the maternal gene, or from a *GY* plant and receive the paternal gene. Since this is an "or" expression joining some of the outcomes of a permissible set we can use Property 3 and add the individual probabilities.

$$\begin{aligned} P_0(Y) &= P_0(YY)L + P_0(YY)R + P_0(YG)L + P_0(GY)R \\ &= (0.10)(1/2) + (0.10)(1/2) + (0.25)(1/2) + (0.25)(1/2) \\ &= 0.35. \end{aligned}$$

---

***Find* $P_0(G)$*, the probability that a gamete of the parental generation will carry the* G *gene.***

---

This calculation is similar to the previous one and we should get that $P_0(G) = 0.65$. The distribution of types for the gametes is 0.35 for yellow and 0.65 for green.

---

***Find* $\mathbf{P_1(YY)}$, $\mathbf{P_1(YG)}$, $\mathbf{P_1(GY)}$, *and* $\mathbf{P_1(GG)}$, *the distribution of types for the first filial generation.***

---

In order for a seed to be *YY*, it must come from an egg cell containing the *Y* factor and a pollen cell carrying the *Y* factor. The probability that an egg cell contains the *Y* factor is $P_0(Y) = 0.35$. The probability that a pollen cell carries the *Y* factor is the same. In random pollination which egg cell and which pollen cell unite is a matter of chance; that is, the members of the pair are independent and Property 4 can be used, so

$P_1(YY) = P_0(Y) \cdot P_0(Y) = (0.35)(0.35) = 0.12$. Similarly,
$P_1(YG) = P_0(Y) \cdot P_0(G) = (0.35)(0.65) = 0.23$,
$P_1(GY) = P_0(G) \cdot P_0(Y) = (0.65)(0.35) = 0.23$, and
$P_1(GG) = P_0(G) \cdot P_0(G) = (0.65)(0.65) = 0.42$.

Now we would like to know what happens in the second filial generation.

---

***Find* $\mathbf{P_1(Y)}$ *and* $\mathbf{P_1(G)}$, *the distribution of types of the gametes produced by the first filial generation.***

---

This calculation is similar to the one for the gametes of the parental generation.

$$\begin{aligned}
P_1(Y) &= P_1(YY)L + P_1(YY)R + P_1(YG)L + P_1(GY)R \\
&= (0.12)(1/2) + (0.12)(1/2) + (0.23)(1/2) + (0.23)(1/2) \\
&= 0.35, \text{ and} \\
P_1(G) &= P_1(GG)L + P_1(GG)R + P_1(GY)L + P_1(YG)R \\
&= (0.42)(1/2) + (0.42)(1/2) + (0.23)(1/2) + (0.23)(1/2) \\
&= 0.65.
\end{aligned}$$

---

***What is the distribution of types for the second filial generation?***

---

Since $P_0(Y) = P_1(Y)$ and $P_0(G) = P_1(G)$, we do not have to carry out the calculation. The distribution of plants in the second filial generation will be exactly the same as the first! In fact, no matter how many generations of random pollination we follow, 12% of the plants will be *YY*, 23% will be *YG*, 23% will be *GY*, and 42% will be *GG*.

This strange result which we calculated for a specific garden of pea plants also occurs when we follow a single characteristic in any form of organic life. *Under random mating the distribution of types of gametes is constant in each generation and the distribution of types for the offspring is constant from the first filial generation on.* This explains why national characteristics persist in a population when there is little outside marriage. Each generation has the same distribution of characteristics as the previous one. If 90% of the population has brown hair in a generation after the population has stabilized, then 90% of the next generation will have brown hair.

The example we followed above concerning pea plants dealt with a garden with a specific number of certain types of plants. In order to convince ourselves that the same thing happens in any population, we must do the calculation in general terms.

Let $A$ and $B$ represent two forms of a characteristic (as the yellow and green seeds). Let $P_0(AA)$, $P_0(AB)$, $P_0(BA)$, $P_0(BB)$ represent the distribution of the parental generation (as above, we assume $P_0(AB) = P_0(BA)$). The probability of a gamete receiving the maternal gene is $L = 1/2$ and of receiving the paternal gene is $R = 1/2$. The distribution of the gametes will be represented by $P_0(A)$ and $P_0(B)$. Similarly, $P_1$ and $P_2$ will be used for the probabilities in the first and second filial generation.

THEOREM: Under random mating the distribution of the gametes of the parental and the first filial generation are the same, and the distribution of types in the first filial generation is the same as in the second. In symbols, if $X$ is one of $A$ or $B$ and $Y$ is one of $A$ or $B$, then $P_0(X) = P_1(X)$, and $P_1(XY) = P_2(XY)$.

*Proof:* We can find the gamete distribution as we did for the specific case before

$$P_0(X) = P_0(XA)L + P_0(AX)R + P_0(XB)L + P_0(BX)R.$$

Since $P_0(XY) = P_0(YX)$ in all cases, we can make a substitution,

$$\begin{aligned} P_0(X) &= P_0(XA)L + P_0(XA)R + P_0(XB)L + P_0(XB)R \\ &= P_0(XA)(L + R) + P_0(XB)(L + R) \\ &= P_0(XA)(1/2 + 1/2) + P_0(XB)(1/2 + 1/2) \\ &= P_0(XA) + P_0(XB). \qquad [1] \end{aligned}$$

In particular, the probability of a gamete carrying the $A$ gene is

$$P_0(A) = P_0(AA) + P_0(AB)$$

and the probability of a gamete carrying the $B$ gene is

$$P_0(B) = P_0(BA) + P_0(BB).$$

We should notice that $P_0(A) + P_0(B) = 1.$ [2]

Using this gametic distribution we can find the distribution of types in the first filial generation. The single formula

$$P_1(XY) = P_0(X) \cdot P_0(Y) \qquad [3]$$

will represent the probability of each type.

The gametic distribution from this first filial generation can be found as above, so is similar to [1]

$$P_1(X) = P_1(XA) + P_1(XB).$$

Substituting [3] into this last expression we get

$$\begin{aligned} P_1(X) &= P_0(X) \cdot P_0(A) + P_0(X) \cdot P_0(B) \\ &= P_0(X)[P_0(A) + P_0(B)]. \end{aligned}$$

Then using [2] we can substitute 1 for the quantity in square brackets, and

$$P_1(X) = P_0(X). \qquad [4]$$

This is the first property we were to prove.

Finally, $P_2(XY) = P_1(X) \cdot P_1(Y) = P_0(X) \cdot P_0(Y)$ from [4], so

$$P_2(XY) = P_1(XY) \text{ from [3]}. \blacksquare$$

## Exercises

**1.** What is the distribution of gametes which would be produced by the following plants?

(a) 20% pure yellow seeds, 10% pure green, 70% hybrid yellow-green.
(b) 20% pure yellow, 80% pure green.
(c) 100% hybrid yellow-green.

**2.** Find the distribution of types for the first filial generation after random mating of each of the parental generations given in Exercise 1.

**3.** If a population of 15% pure blue-eyed people, 21% pure brown-eyed people, and 64% brown-eyed people with a recessive gene for blue intermarry, what will be the distribution of genes for eye color in the next generation? In the generation after that, if there is no outside marriage? (Assume that eye color is not an important factor in the selection of a marriage partner and that all couples have essentially the same number of children.)

### 6. WHAT HAVE WE LEARNED FROM THE PROBABILITY OF GENETICS?

There is an important difference between this chapter and all of the preceding ones in this book. All of the others were mainly about *pure* mathematics. This one has dealt with an example of *applied* mathematics.

It is true that in the other chapters we often noticed that some concrete situation was the basis of a mathematical abstraction. We also noticed that after a part of mathematics was developed at the abstract level it often had many applications in the real world. Our attention, however, was always focused on the abstract mathematical development. We were looking at pure mathematics.

In this chapter we worked up to Hardy's theorem. For mathematics, the facts reached in the conclusion are not of great value; however, when they are applied to the real world, the conclusion is one of the outstanding discoveries of this century. It was reached, not by experimentation, but by reasoning from Mendel's First Law and the principles of probability. (After Hardy published his result in *Science* in 1908, it was experimentally verified by biologists.) Hardy's work is a good example of applied mathematics.

If Hardy were alive today he would probably protest being used as an example of someone who did applied mathematics. Hardy above all typifies the attitude of a pure mathematician (see his *A Mathematician's Apology).* For him, *beautiful* rather than *useful* was the first test of valuable mathematics. He even said, "I have never done anything 'useful,'" and "I have . . . very little interest in experimental science." Hardy knew that he had an unusual gift for doing pure mathematics and dedicated his life to it. Almost in spite of himself, he got interested in genetics at least long enough to write one paper about it, a very important paper at that.

Other mathematicians are constantly aware of applications. They often develop some pure mathematics, but are motivated by the use it will have in one or more of the sciences (Wiener's book and Kline's article, listed at the end of this section, indicate this attitude). It is senseless to get into debates about whether mathematics is basically pure or basically applied, or about which variety is more important. Both beauty and usefulness have value. In the case of mathematics, history has shown that both of these elements have flourished because of the contributions they have made to each other.

Applied mathematics appears in every science—physics, chemistry, biology, medicine, economics, the social sciences, and linguistics are some of the broad areas. In each case, the role of mathematics is similar to the role it played in our study of genetics. Some part of mathematics is found which mirrors the physical properties being studied. (The mathematics may exist before the study starts, or the physical problem may motivate the development of some new mathematics.) The language and methods of mathematics are then used to deduce new information about the physical phenomenon.

## Exercises

There are many expository articles on applied mathematics which have been written for nonmathematicians. Most of them have been written by people who are very concerned about communicating what they are doing to the average person. Read one, or more, of the following:

Cohen, Hirsch, "Mathematics and the Biomedical Sciences," *The Mathematical Sciences,* Edited by COSRIMS, M.I.T. Press, Cambridge, Mass., 1969, pp. 217–231.

*Dalton, A. G., "The Practice of Quality Control," *Scientific American,* Vol. 188, No. 3, March, 1953, pp. 29–33.

*Dirac, P. A. M., "The Evolution of the Physicist's Picture of Nature," *Scientific American,* Vol. 208, No. 5, May, 1963, pp. 45–53.

*Dyson, Freeman, J., "Mathematics in the Physical Sciences," *Scientific American,* Vol. 211, No. 3, September, 1964, pp. 128–146.

Hardy, G. H., *A Mathematician's Apology,* Cambridge University Press, Cambridge, 1941. (An excerpt is in *The World of Mathematics,* Vol. 4, J. R. Newman, Simon and Schuster, New York, 1956, pp. 2027–2038.)

Kemeny, John G., "The Social Sciences Call on Mathematics," *The Mathematical Sciences,* Edited by COSRIMS, M.I.T. Press, Cambridge, Mass., 1969, pp. 21–36.

Klein, Lawrence R., "The Role of Mathematics in Economics," *The Mathematical Sciences,* Edited by COSRIMS, M.I.T. Press, Cambridge, Mass., 1969, pp. 161–175.

Kline, Morris, "The Import of Mathematics," *Mathematics in the Modern World,* Readings from Scientific American, W. H. Freeman, San Francisco, 1968, pp. 232–237.

*Moore, Edward F., "Mathematics in the Biological Sciences," *Scientific American,* Vol. 211, No. 3, September, 1964, pp. 148–164.

Salvadori, Mario G., "Mathematics, The Language of Science," *An Outline of Man's Knowledge of the Modern World,* Edited by Lyman Bryson, Doubleday, Garden City, N. Y., 1960, pp. 189–209.

*Stone, Richard, "Mathematics in the Social Sciences," *Scientific American,* Vol. 211, No. 3, September, 1964, pp. 168–182.

Sutton, O. G., *Mathematics in Action,* Thomas Y. Crowell, New York, 1955.

Wiener, Norbert, *I Am a Mathematician,* M.I.T. Press, Cambridge, Mass., 1956.

Wigner, Eugene, "The Unreasonable Effectiveness of Mathematics in the Natural Sciences," *Communications on Pure and Applied Mathematics,* Vol. 13, No. 1, February, 1960, pp. 1–14.

Ulam, Stanislaw, "The Applicability of Mathematics," *The Mathematical Sciences,* Edited by COSRIMS, M.I.T. Press, Cambridge, Mass., 1969, pp. 1–6.

Also see the references in Section 5, Chapter 5. They, too, deal with applied mathematics.

## 7. WHAT IS PROBABILITY?

Although we have used this chapter on probability to illustrate methods of applied mathematics, we should not get the impression that probability is never studied in an abstract way for its own sake. Today, abstract mathematical probability is a highly respected branch of mathematics, in which some axioms for probability that look much like the principles we mentioned in Section 4 are assumed. The undefined objects of the axiom system are events (or outcomes), and the probability of an event is an undefined relation (between an event and a real number) which satisfies the axioms.

This general approach to probability has much in common with measure theory—a generalization of the length of a segment. We do not want to go into detail, but will indicate some of the things which are similar. Assume that we have a line segment $U$ of length 1, we'll write $L(U) = 1$. It is reasonable to extend the notion of length to a single point; it has length 0. If $A$ is any segment contained in $U$, then $0 \leq L(A) \leq 1$; if $A$ is a single point $L(A) = 0$: if $A = U$, $L(A) = 1$. (This looks very much like Principle 1.) It is also reasonable to extend the notion of length to a union of disjoint segments. If $A$ and $B$ are two segments in $U$ which do not overlap and $A \cup B$ is the union of $A$ and $B$, then $L(A \cup B) = L(A) + L(B)$. (This is similar to Principle 3.) Although probability theory is much like measure theory, it is an independent mathematical discipline, whose problems would be of little interest to someone developing measure theory.

Now that we know something of the mathematical nature of probability, it seems appropriate to end this chapter with a few warnings about how we should apply the results of probability to our lives.

One frequent mistake was mentioned in Section 2. The predictions of probability apply to a collection, and *not* to the individuals of a collection. The probability that a driver will have an accident sometime in his life is 1/4, but this tells us nothing about whether we will have an accident. Similarly, the information that the probability of having a serious side effect from a certain drug is 0.05 tells us absolutely nothing about what an individual's reaction to the drug will be.

In practice, of course, we often must make decisions on the basis of information that probability gives us. The probability of 1/4 for being involved in an accident does not keep us from driving. If a doctor feels that the drug with probable danger of 0.05 is the best thing available to cure a patient, he will prescribe it. Whether we are in an accident, or whether the patient will have a serious reaction will only be known after the fact. Since we do not have certain information about most things, this is the best we can do.

Another frequent misinterpretation concerns independent events. The probability that two hybrid blue-brown eyed people will have a brown-eyed child is 3/4. If Mr. and Mrs. Jones meet this description and already have three brown-eyed children, they feel that the probability that their fourth child will be blue-eyed is very high—so that they will fit the three to one ratio. This is not what probability theory is telling them. It says that if we consider a collection of many couples like themselves and all of their children, then approximately three out of four children will be brown-eyed. It says nothing about the individual family. The birth of each child in a family is independent of the others, so the chance of Mr. and Mrs. Jones having a blue-eyed fourth child is the same as the probability of having a blue-eyed first child—1/4.

A similar misconception leads some people to bet highly on tossing a head after a string of several tails. They feel that the chance of the string being broken is very great because in the long run heads must appear 1/2 of the time. Actually, each toss is independent, and the string of tails has no influence on the next toss. (A string of tails, on the contrary, should lead them to suspect that the coin is biased in favor of tails, and if they do not bet evenly, then it should be in favor of tails.)

Probability theory touches our lives constantly. Aside from the misuses of probability, even correct use should carry a warning. Probability affects insurance premiums, availability of telephone lines, calculation of I. Q.'s, permission to retail drugs, weather forecasting, campaigns against cigarette advertising, quality control of products, decisions to undergo surgery, and hopes for a favorite team. The information that probability theory brings to these areas is very valuable; however, we must remember that it does not answer everything. If we reduce a problem to numbers and science, we sometimes tend to feel that everything is tidy and definite. Nothing is further from the truth—at best, probability is a measure of uncertainty.

## Exercises

### READ FROM THE FOLLOWING

*Ayers, A. J., "Chance," *Scientific American,* Vol. 213, No. 4, October, 1965, pp. 44–54.

Bergamini, David, and the Editors of Life, *Mathematics,* Life Science Library, Time Incorporated, New York, 1963, pp. 126–147.

*Kac, Mark, "Probability," *Scientific American,* Vol. 211, No. 3, September, 1964, pp. 92–108.

*Weaver, Warren, "Probability," *Scientific American,* Vol. 183, No. 4, October, 1950, pp. 44–47.

The article on Probability, Mathematical in a recent edition of the *Encyclopaedia Britannica.*

## Test

1. What is the probability of tossing each of the following on a single toss of one die?

   (a) An even number
   (b) The number 5
   (c) A number different from 5
   (d) The number 7
   (e) A number different from 7
   (f) An even number or a number greater than 3
   (g) An even number and a number greater than 3

2. When tossing a single die is {an even number, a number greater than 3} a permissible set of outcomes? Why?

3. Are the outcomes *an even number* and *a number greater than 3* independent? Why?

4. Brown eyes are dominant over blue eyes.

   (a) If a genetically pure brown-eyed person marries a genetically pure blue-eyed person, what is the probability that if they have a child he will be brown-eyed?
   (b) If the couple in part (a) have a brown-eyed child and he marries a genetically pure brown-eyed person, what is the probability of them having a blue-eyed child? A brown-eyed child with a recessive gene for blue?
   (c) If the couple in part (a) have a brown-eyed child and he marries a brown-eyed person with a recessive gene for blue, what is the probability of them having a blue-eyed child?
   (d) If the couple in part (c) have six blue-eyed children in a row, what is the probability that their seventh child will be blue-eyed?

5. Let a nickel with $P(H) = 0.30$ and a dime with $P(t) = 0.60$ be tossed. Find:

   (a) $P(T)$ if $T$ stands for the tail of the nickel
   (b) $P(h)$ if $h$ stands for the head of the dime
   (c) $P(Hh)$
   (d) $P(Tt \text{ or } Hh)$
   (e) $P(\text{coins have different faces up})$

6. If a population of 20% genetically pure blue-eyed people, 30% genetically pure brown-eyed people, and 50% brown-eyed people with a recessive gene for blue intermarry, what will be the distribution of genes for eye color in the next generation? In the generation after that if there is no outside marriage?

7. What outside reading did you do about probability or applied mathematics? Summarize anything additional to the text which you have learned.

IBM

9

# COMPUTER SCIENCE

## Eliminating the Tedious

**Education in our society is incomplete without some knowledge of computers.**

Many people would question whether computer science is a branch of mathematics. Computer science is the study of computers, and computers are exactly what the word implies—machines which can perform computations. Computations are usually not the main interest of mathematicians. Paradoxically, computers are making significant contributions to pure mathematics. A look at this relationship between mathematics and computers will contribute to our attempt to understand the nature of mathematics.

Besides aiding mathematicians, computers are playing an ever-increasing role in our lives. Computers are used to schedule college classes, help locate brain tumors, predict the weather, manage credit card accounts, assist in translating languages, check income tax reports, prepare payrolls, assist in crime prevention, locate information, and, of course, do innumerable computations for engineers and scientists. An education today is incomplete without some knowledge of how this is all possible.

### 1. WHAT IS A COMPUTER?

A well-known mathematician once said, "One of the prerequisites of a good mathematician is laziness." Perhaps this explains why the inventors of some of the mechanical devices which were forerunners of the present-day computers were giants in mathematics.

The first successful mechanical adding and subtracting machine was invented by Blaise Pascal. He began planning it in 1642, at the age of 19, because he was tired of the many computations he had to do for his tax-collector father. In 1674, Gottfried Leibniz, one of the independent inventors of calculus (see Section 7, Chapter 7), built an all-purpose machine which could multiply, divide, and find square roots, as well as add and subtract. The direct descendants of these machines are the present-day adding machines and desk calculators.

Modern *digital computers* are like the inventions of Pascal and Leibniz in that they are basically machines which perform arithmetic operations. In both a desk calculator and a computer we find the following sequence:

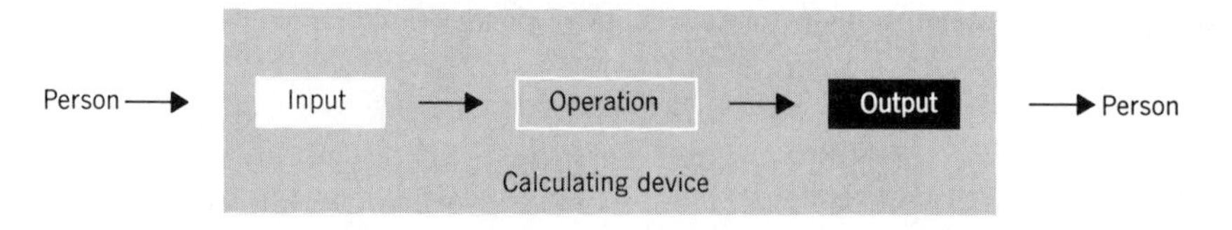

The input and output are the "communication" process between man and machine. A person gives the machine certain data (numbers) and instructions. If he is working with a desk calculator, the input may be accomplished by pushing certain keys. In a computer, the data and instructions may enter the machine in the form of punched cards, paper tape, or magnetic tape. In both cases the machine carries out the instructed operations on the data and then exhibits the result in some way that can be perceived by the operator. A desk calculator may print this output on paper or display the number in a register. A computer may punch the result on paper tape or cards, or write it on magnetic tape and then print it out.

Theoretically, there is nothing that a computer does that can't be done on a desk calculator, given enough paper for all the data, enough logical organization, enough clerical help, and enough time and patience. In practice, however, it would take years to accomplish what the computer is able to do in minutes (some computers do over a million computations a second). This is the source of their great value.

Besides the speed of a computer, there are some other essential differences between a conventional desk calculator and a digital computer. In the case of the desk calculator the data are kept outside of the machine and entered one step at a time by the operator. Similarly, the list of instructions remains outside the machine (on a sheet of paper or in the head of the operator) and only one instruction is given the machine at a time. The digital computer is given all the data and instructions at the outset. These are stored in the "memory" of the computer until needed. Another difference is the ability of the computer to make decisions between different possible operations, depending on the result of a previous operation. For the desk calculator it is the operator who must make such decisions.

The first plans for building a digital computer were made by Charles Babbage (1792–1871). He called it the Analytic Engine. It was designed so that the data would be introduced by punched cards, it would remember up to 1000 fifty-digit numbers, it could carry out any arithmetic operation, the instructions would be built in, and the instructions could be modified by the result of a previous computation. This self-regulating feature led Babbage to describe his machine as "eating its own tail." Babbage's elaborate plans were far ahead of his time. The machine envisioned was completely mechanical (electronic equipment did not exist) and involved about 50,000 parts. Because precision tooling was not the highly developed art it is today, the Analytic Engine was never successfully completed. Babbage's dream was not to materialize until the 1940s and the invention of electronic digital computers.

Babbage would be gratified to see how similar in organization a modern computer is to his. The specific details, however, are quite different and involve extremely complex electronic equipment. One innovation that made all of this possible was the switch to binary notation (see Section 4, Chapter 6) from our usual decimal notation for numbers. Since binary notation uses only two symbols, (1, 0) it is easy to transfer numbers written in this notation to punched cards (hole, no hole), magnetic tape, drum, or disk (magnetic spot, no magnetic spot), electrical switches (current, no current), or core magnets (clockwise field, counterclockwise field).

The memory of a computer utilizes some form of magnetic recording. For fast retrieval, units of tiny core magnets, magnetic drums or disks, or tiny magnetic spots on film are used. For permanent information which is not frequently used, or when speed is not an object, magnetic tape can be used in the memory. Babbage had plans for a memory of 1000 numbers. Some modern computers can store 10,000,000,000 words (numbers or instructions) with quick accessibility. If speed is not a factor, the storage on magnetic tape is almost unlimited.

The parts of the computer that control the entire process and actually perform the arithmetic were to be done by wheels, gears, and levers in Babbage's machine. The first modern computer used electrical relays. These were not always reliable and so were later replaced by vacuum tubes. Vacuum tubes, however, are costly to operate; they require much power and create a great deal of heat which must be cooled. Today transistors have taken over the job, and computers are smaller, less costly, more reliable, and less expensive to operate.

Besides the digital computers we have been discussing, another type of electronic computer, called an *analog computer,* is in use today. Digital and analog computers are built on completely different principles. A digital computer reduces the arithmetic operations to counting. In an analog computer, quantities are represented by electrical quantities, and amplifiers and capacitors are used to change this quantity in a manner *analogous* to the various mathematical operations.

The difference between these two approaches is illustrated by the following example. We could count a stack of pennies (a digital approach) or we could measure the height of the stack and, knowing that 16 pennies measure one inch, we could figure out how many pennies are in the stack (an analog approach). In this second approach we get a measurement, but the analogous information we are interested in is the number of pennies. Analog devices are common. A mercury thermometer is an analog device. We measure the length of the mercury in a tube, but the information we want is temperature.

Since analog and digital computers are so different, it is not surprising that each one is better than the other for certain jobs. For some jobs analog computers are faster because they can do several operations simultaneously; the digital computer must perform one operation at a time. On the other hand, computers which work by analogy are not always as accurate as digital computers because they depend on the accuracy of the measurement. Recently, some hybrid computers have been developed which make use of the best features of each system.

The remainder of this chapter will deal with digital computers. We will look at how a problem must be organized to prepare it for a computer. We will see that some of our familiar arithmetic techniques can be revised to better fit the computer. We will then mention some important advances in mathematics that have resulted from computers, look at some non-mathematical uses of computers, and speculate a little about the future of this invention which is changing our lives so rapidly.

## Exercises

**1.** In each of the following situations indicate the input, operation, and output.

(a) using a typewriter
(b) asking a person to answer a question verbally
(c) using a dial telephone
(d) using an alarm clock
(e) operating a soft drink machine
(f) opening a combination lock
(g) turning on a TV set
(h) using a mercury thermometer

**2.** In each of the following analog devices, identify the physical quantity which is measured and the analogous desired information.

(a) a spring scale
(b) gasoline gauge
(c) a barometer
(d) a speedometer
(e) a slide rule
(f) an hourglass

**3.** Indicate which of the following devices are of an analog nature and which are digital.

(a) a protractor
(b) a sundial
(c) tallying votes at an election
(d) a cash register
(e) a ruler
(f) an abacus
(g) a clock
(h) a counter used to tally attendance at an event
(i) poker chips
(j) an adding machine

**4.** Read one or more of the following:

Bergamini, David, and the Editors of Life, *Mathematics,* Life Science Library, Time Incorporated, New York, 1963, pp. 18–37.

Bernstein, Jeremy, *The Analytical Engine: Computers—Past, Present, and Future,* Random House, New York, 1964.

Bronowski, Jacob, "Science as Foresight," *What is Science?* Edited by James R. Newman, Simon and Schuster, New York, 1955, pp. 385–436.

*Davis, Harry M., "Mathematical Machines," *Scientific American,* Vol. 180, No. 4, April, 1949, pp. 28–39.

*Evans, David C., "Computer Logic and Memory," *Scientific American,* Vol. 215, No. 3, September, 1966, pp. 74–85.

*Morrison, Philip and Emily, "The Strange Life of Charles Babbage," *Scientific American,* Vol. 186, No. 4, April, 1952, pp. 66–73.

Richtmyer, R. D., "The Post-War Computer Development," *The American Mathematical Monthly,* Vol. 72, No. 2, Part II, February, 1965, pp. 8–14.

Rosenberg, Jerry M., *The Computer Prophets,* Collier-Macmillan, Toronto, 1969.

Sutherland, Ivan E., "Computer Inputs and Outputs," *Scientific American,* Vol. 215, No. 3, September, 1966, pp. 86–96.

The article on Computer in a recent edition of the *Encyclopaedia Britannica.*

## 2. PROGRAMMING—FLOW DIAGRAMS AND CODING

So far we have been talking about the computer itself (the hardware) and not the instructions and data (the software) that are given to the computer. The process of preparing information for a computer is called *programming*. There are two aspects of programming. The first is to analyze the problem to be solved and to organize it in a logical step-by-step manner. This is called *systems analysis* or making a *flow diagram.* The second is to translate these steps into *machine language* so that it can be "understood" by the computer.

The necessity of organizing a process was dramatically brought out by a highly acclaimed National Educational Television program for preschool children—"Sesame Street." Daily, "Sesame Street" featured a skit between Buddy and Jim which revolved around the almost slapstick situations which occur when organization is not part of problem solving. Buddy, for example, was faced with the problem of putting on his shoes. At first he put his shoes on, then his socks. A second attempt got the shoes on the wrong feet. The third attempt resulted in the shoes on the right feet, but tied together. The lesson they were trying to teach to preschoolers is the same one that a programmer must first learn: Before any problem can be solved we must decide which steps to take, in which order to take them, and which decisions must be made and when.

One way to help organize a problem is to construct a *flow diagram.* Buddy would have had no problem if he had the self-explanatory diagram in Fig. 9–1.

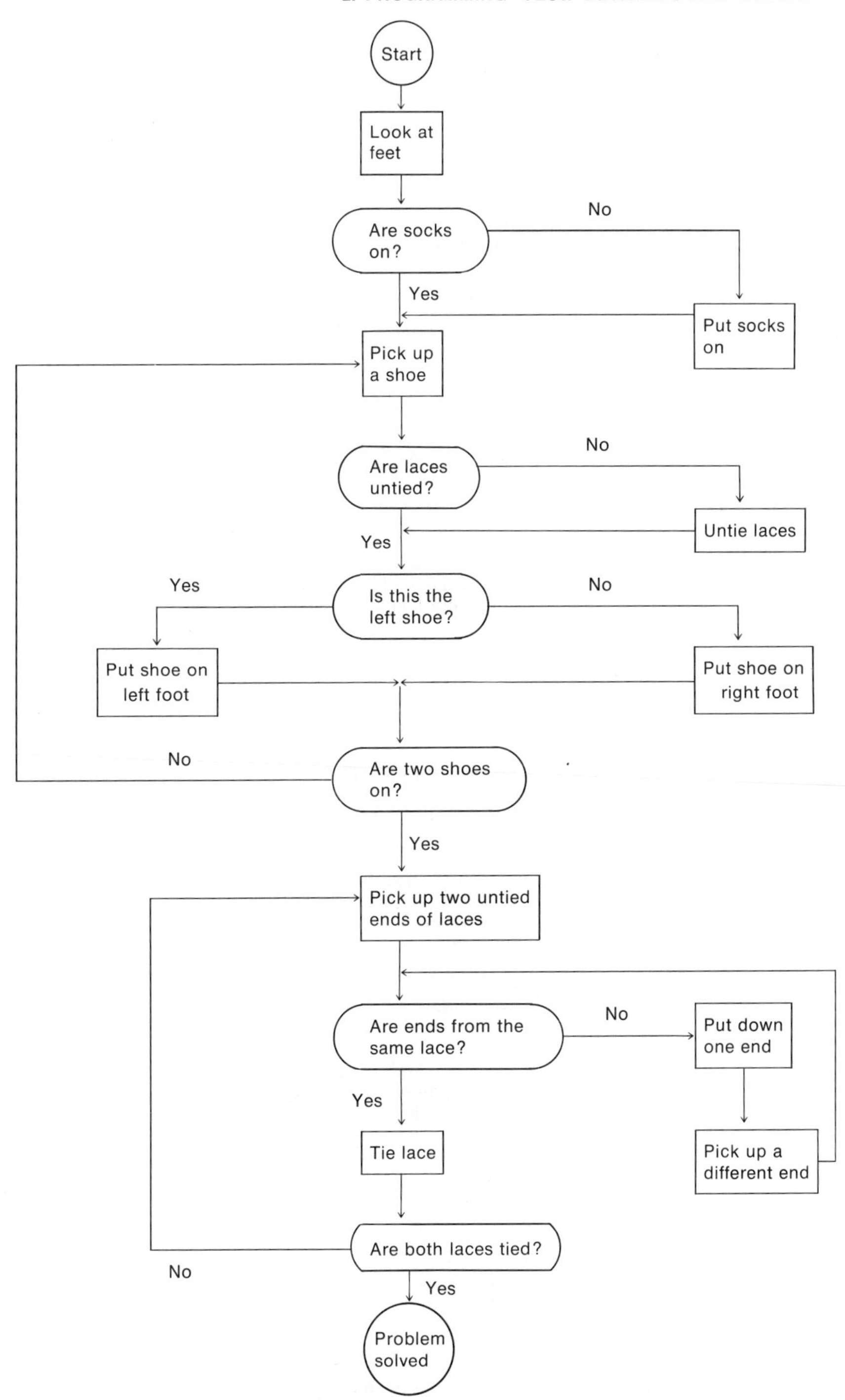


**Fig. 9–1. Putting on shoes with laces.**

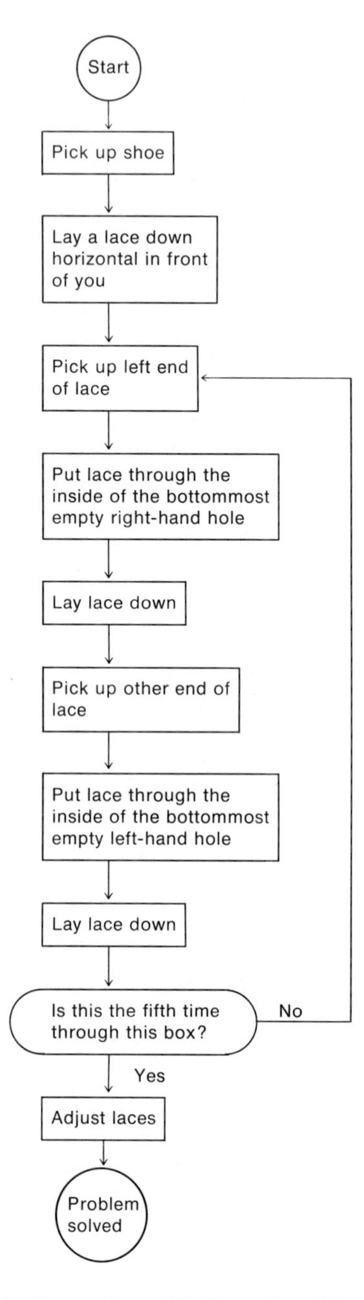


**Fig. 9–2. Lacing a shoe with five pairs of holes.**

Notice that the rectangular boxes in the flow diagram represent actions to be taken, the ovals are decisions, and the circles are the start and finish. Flow diagrams similar to this one are actually used by programmers in the first stages of preparing material for a computer.

***Construct a flow diagram for lacing a shoe with five pairs of holes.***

There is, of course, more than one solution to a problem like this one. There is a certain satisfaction in being able to find a short direct approach to a problem. In the case of preparing a computer program, the simpler the flow diagram the easier it will be to carry out the second stage of translating the problem into machine language. One solution to the above problem is shown in Fig. 9–2.

This flow diagram could also contain provisions for keeping the length of the two sides of the lace even, what to do in case of a break or knots, and so on. It might be a challenge to enlarge this diagram to include some of these contingencies. The thing we should notice about our solution is that we kept the diagram relatively short by using a decision making step, "Is this the fifth time through this box?" It was not necessary to write separate instructions for each pair of holes. Of course, it is necessary to have a way of determining how many times the box is passed through. One way to do this is to keep a tally. The tallying procedure could also be included in the diagram if desired. Flow diagrams can be as detailed as we desire. The more detail, the closer they are to the way a computer would actually be instructed.

A flow diagram with several decisions can get more complex. Figure 9–3 shows a flow diagram for baking a cake. Notice that it does not contain explicitly all of the steps because the action "mix the cake as directed in the recipe" refers us to an already existing flow diagram—the recipe. This illustrates that often a complicated problem can be simplified by making use of parts that are already solved. Situations similar to this occur in actual programming. The parts are called *subroutines.*

The task of organizing a mathematical problem is much the same as the preceding examples. The flow diagram for the following problem appears in Fig. 9–4.

***Construct a flow diagram for subtracting two-digit natural numbers with a provision for borrowing on a computer which can handle single-digit subtraction.***

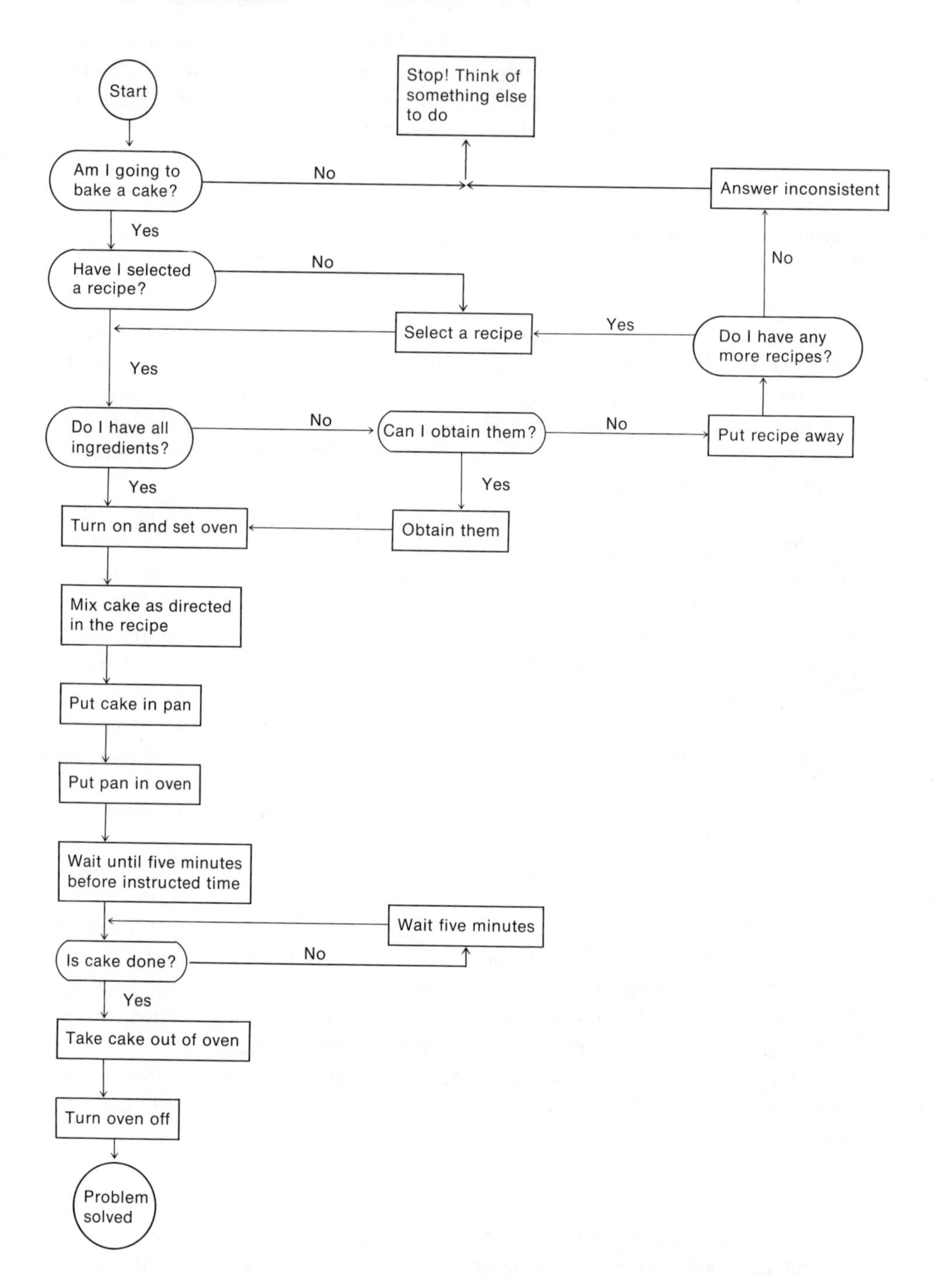


**Fig. 9–3. Baking a cake.**

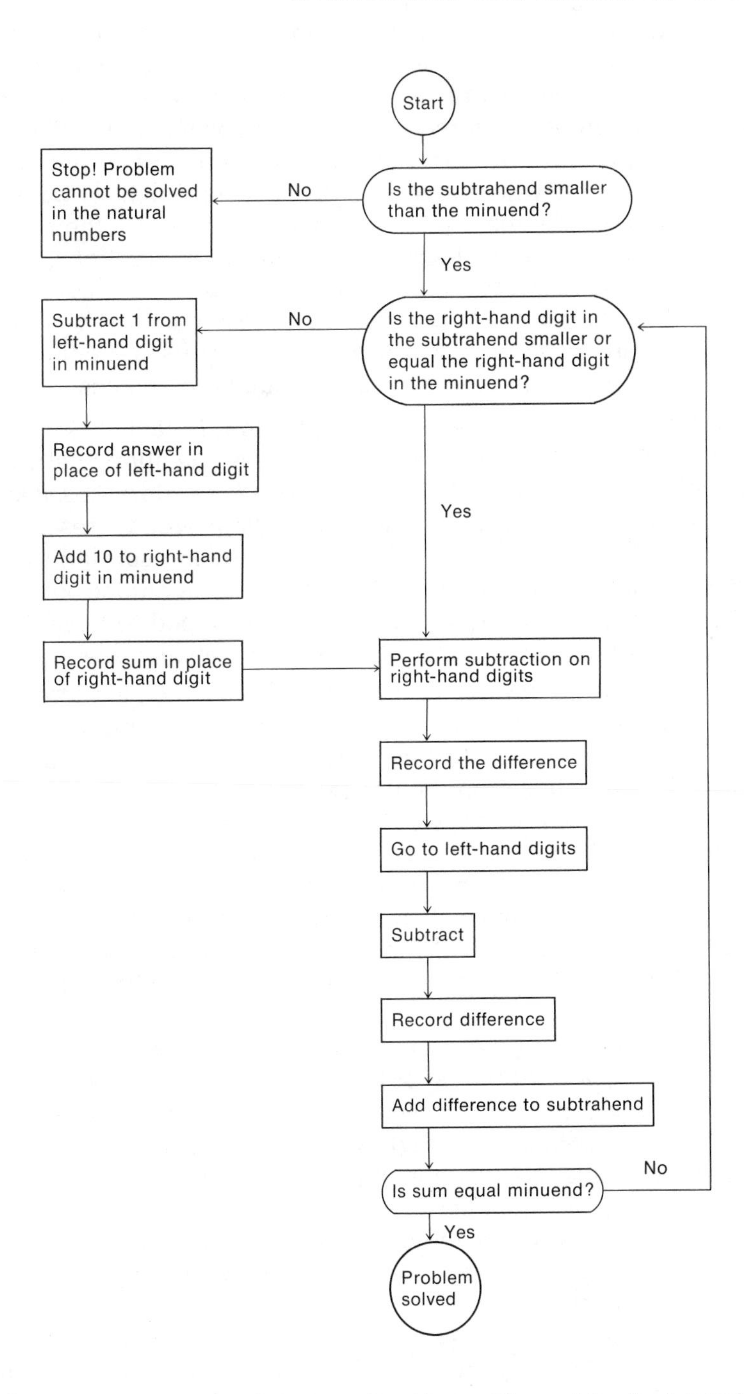


**Fig. 9–4. Subtracting two-digit natural numbers with provision for borrowing.**

The second stage of programming, translating the flow diagram into machine language (coding), was a job formerly left entirely to professional programmers. Since the computer is made to receive only numbers, all of the instructions are entered by means of code numbers. When letters are in the data, these too are represented by code numbers. The computer must be told precisely where to store the numbers which are data, where to store the numbers which are instructions, where to put the result, and when to print out the result. To avoid long lists of instructions, repetitive loops (like the one that occurred in the flow diagram for lacing a shoe) are used and if necessary a tally is made and stored in the computer. The storage places are given numbers, too, called *addresses.* Coding, then, is concerned with storing numbers in certain numbered places.

This coding process can become complicated. In fact, it is usually extremely complicated! It is necessary to remember the numerical codes for the arithmetic operations and to keep track of all the addresses. It is sometimes necessary to resort to complicated tallying procedures. Also, it is difficult to modify a list of instructions once they are set up. If a step is missing or is unnecessary the entire program must be redone. Fortunately, since the late 1950s techniques have been developed to have the computer do much of its own translating into machine language.

The solution was to develop several languages which are intermediate between English (or some other natural language) and the machine language. These programming languages contain the technical phrases of the discipline which is going to use the computer, and the computer is then programmed to turn these phrases into machine language. This translating program is called a *compiler.* One phrase in the program language may be changed into several machine language instructions (sometimes as many as twenty) by the compiler. Some of the program languages used are COBOL (Common Business Oriented Language), FORTRAN (Formula Translation), and ALGOL (Algorithmic Language). There are many more.

The use of these programming languages has made it possible for many people to use a computer without employing a professional programmer constantly. It is necessary to first learn the language, but this can be done in a few weeks in a short course. It is common for librarians, pyschologists, people in education, business, social science, and medical research, as well as engineers, scientists, and mathematicians (who usually are not trained in computer science) to take such a course. They are then able to "communicate" with the computer by themselves for many of their problems.

Programming languages have simplified a great deal of the problem of communication between man and the machine. However, they are extremely rigid (the computer, for example, will reject a command that is not punctuated correctly) and restricted to a small number of technical phrases. The ideal situation would be if we could simply address the computer in English. Some advances have been made along these lines (see the article referred to in Exercise 9 by Bross *et al.*), but at present this is not possible.

## Exercises

**1.** Construct a flow diagram for starting a car with automatic transmission.

**2.** Construct a flow diagram for making coffee in an electric percolator.

**3.** (a) Construct a flow diagram for crossing an intersection with traffic lights and no vehicular traffic.
(b) Enlarge the above diagram for the possibility of vehicular traffic.

**4.** Construct a flow diagram for using a soft drink machine.

**5.** Construct a flow diagram for using a dial telephone.

**6.** Select a "problem" situation of your own and construct a flow diagram of its solution.

**7.** (a) Construct a flow diagram for adding two two-digit natural numbers without carrying.
(b) Expand the above diagram to include the possibility of carrying.

***8.** (a) Construct a flow diagram for changing base-two numerals to base ten.
(b) Construct a flow diagram for changing base-ten numerals to base two.

**9.** Read one of the following:

Bowden, Lord, "The Language of Computers," *American Scientist*, Vol. 58, No. 1, January-February, 1970, pp. 43–53.

Bross, Irwin D. J., *et al.*, "Feasibility of Automated Information Systems in the Users' Natural Language," *American Scientist*, Vol. 57, No. 2, 1969, pp. 193–205.

Goldstein, M., "Computer Languages," *The American Mathematical Monthly*, Vol. 72, No. 2, Part II, February, 1965, pp. 141–146.

*Strachey, Christopher, "System Analysis and Programming," *Scientific American*, Vol. 215, No. 3, September, 1966, pp. 112–124.

## 3. REPETITIVE PROCESSES

In the last section we found the flow diagrams for some common everyday occurrences, such as putting on shoes. When we analyzed these situations the number of steps and decisions required was surprising. We usually carry out these procedures with what seems like little thought. In the light of this experience, it is not surprising that the arithmetic processes we commonly use also contain many steps and decisions. In this section we will look at some common problems in arithmetic and see some approaches, different from our usual ones, that are better suited to a computer.

Computers are able to receive and identify negative and positive numbers as well as zero. Using this ability, and subtraction, it is possible to devise a simple way to find the larger of two numbers (Fig. 9–5).

***Construct a flow diagram for distinguishing whether* A $>$ B *or* B $>$ A, *if* A$\neq$ B.**

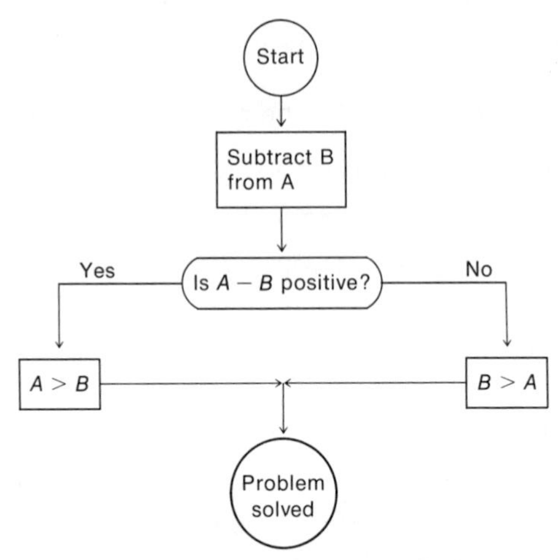


**Fig. 9–5. Determining the larger of two distinct numbers.**

Since it is possible to determine the larger of any two different numbers, it should be possible to find the largest of a finite set of different numbers.

***Construct a flow diagram for finding the largest member of a finite set of distinct numbers.***

Faced with a problem like this, most of us would first see if there were any positive numbers in the set. If there were, then we would drop all the negative numbers and zero (if they were in the list) and consider only the positive numbers. In our memories we have a notion of the numbers in order according to size. Even for the decimal numbers, we have a definite idea of their natural order. Consulting this memory we would decide which positive number in the set is the largest. If there were no positive numbers, we would modify this procedure accordingly.

This approach does not seem particularly adaptable to a computer. It would be better to construct a flow diagram for this problem which does not involve any memory of the numbers in the natural order, a rather complex notion. If it is a good computer program it should be relatively simple and take advantage of repetitive loops. (It also should not use too much computer time, which is expensive, and should give reasonably accurate results.)

The problem of finding the largest member of a finite set can be solved by looking at two numbers at a time, deciding which is larger, and then throwing away the smaller. This process is repeated until only one number is left, the largest. The flow diagram is shown in Fig. 9–6.

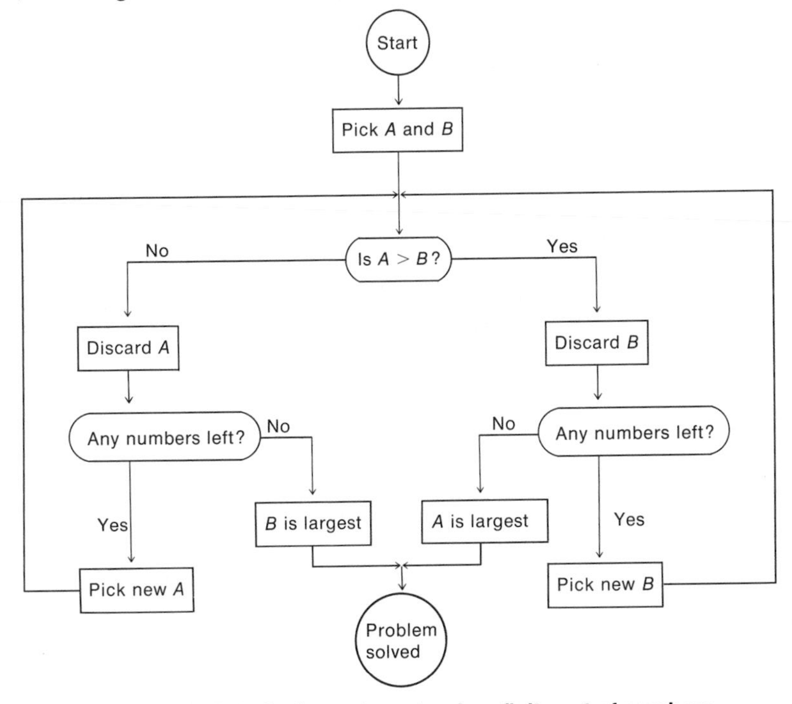


**Fig. 9–6. Finding the largest number in a finite set of numbers.**

A similar thing happens when we prepare other problems for a computer. The approach most natural to us is not always the best for the computer. Multiplication is a good example. To multiply $67 \times 13$ we usually use the following algorithm:

$$\begin{array}{r} 67 \\ \times\ 13 \\ \hline 201 \\ 67\phantom{0} \\ \hline 871 \end{array}$$

This process requires a knowledge of the multiplication table, carrying, shifting position, and addition (sometimes with carrying). It is not necessary to use multiplication to do this problem. Instead the product is the same as the sum of thirteen terms each of which is 67. This requires only addition (with carrying) and keeping a tally (subtraction). See Fig. 9–7.

***Construct a flow diagram for doing multiplication of natural numbers by repeated addition.***

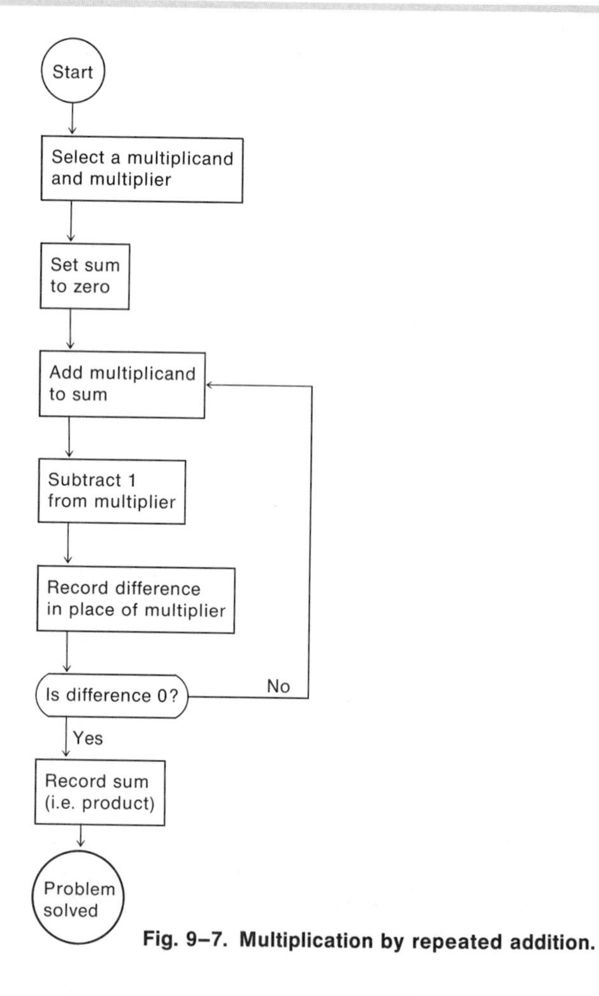


**Fig. 9–7. Multiplication by repeated addition.**

Applying this process to $67 \times 13$, the arithmetic would look as follows:

| Step | Sum | Difference |
|---|---|---|
| 1 | 00 | – |
| 2 | $00 + 67 = 67$ | $13 - 1 = 12$ |
| 3 | $67 + 67 = 134$ | $12 - 1 = 11$ |
| 4 | $134 + 67 = 201$ | $11 - 1 = 10$ |
| 5 | $201 + 67 = 268$ | $10 - 1 = 9$ |
| . | . | . |
| . | . | . |
| . | . | . |
| 14 | $804 + 67 = 871$ | $1 - 1 = 0$ |

***Construct a flow diagram for performing division in the natural numbers by repeated subtraction. (Fig. 9–8)***

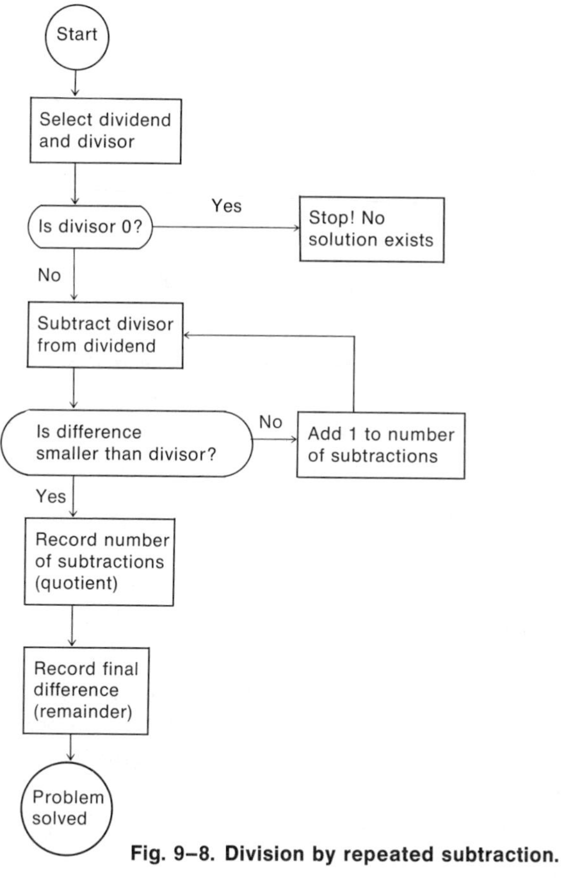


**Fig. 9–8. Division by repeated subtraction.**

## Exercises

1. Use the flow diagram for determining the larger of two numbers to find the larger of $-3$ and 10, $-5$ and $-7$, 0 and 6, 11 and 9, and 0 and $-3$.
2. Revise the first flow diagram in Section 3 to determine whether $A > B$, $B > A$, or $A = B$ for any two numbers $A$ and $B$.
3. Use the flow diagram developed in this section to find the largest number in each of the following sets.
   (a) $\{2, -1, 4, 0, 7, -15, 10, 11, -12\}$
   (b) $\{2/3, 0.01, -15, \pi, -0.25, 36\}$
   (c) $\{3, 15, 6, 31, 7, 10, 43, 2, 17\}$
4. Why was the word "finite" necessary in our consideration of finding the largest number in a set.
5. Perform the following multiplications using repeated addition as described by the flow diagram: $41 \times 8$, $38 \times 7$, $16 \times 10$.
6. Could the flow diagram for multiplication by repeated addition be used for all pairs of integers? If so, why? If not, could some revisions be made? Could it be used for multiplying fractions? Why, or why not?
7. Do the following division problems by repeated subtraction according to the flow diagram in this section.

   (a) $35 \div 5$ (c) $191 \div 21$
   (b) $35 \div 3$ (d) $63 \div 0$
8. What would happen if the decision "is the divisor 0?" had been omitted from the flow diagram for division?
9. Explain why division by repeated subtraction does not work for all integers the way the flow diagram is constructed. Revise the diagram so that division is possible for all integers (except 0 divisors, of course).

*10. Assume that a list of primes in sequence according to size is available. Construct a flow diagram for finding the prime factorization of a natural number.

### 4. ACCURACY

In Section 3 we found simple and repetitive ways to solve some arithmetic problems. It could happen that a simple, repetitive approach is still not the best one to use in a computer. Besides simplicity, there are other factors to consider. Two of these are computer-time and accuracy. Because computer-time is expensive, a slightly more complicated but much shorter procedure may be desired. For example, to multiply $3212 \times 2231$ would require 2232 steps by repeated addition. This problem can still be done by repeated addition but shortened considerably if we make use of a shift in the position of the digits (an operation a computer can perform). The following shows how this problem can be done in 9 steps.

| Step | Sum | Difference |
|---|---|---|
| 1 | 0000 | — |
| 2 | 0000 + 3212 = 3212 | 2231 − 1 = 2230 |
| 3 | 3212 + 32120 = 35332 | 2230 − 10 = 2220 |
| 4 | 35332 + 32120 = 67452 | 2220 − 10 = 2210 |
| 5 | 67452 + 32120 = 99572 | 2210 − 10 = 2200 |
| 6 | 99572 + 321200 = 420772 | 2200 − 100 = 2100 |
| 7 | 420772 + 321200 = 741972 | 2100 − 100 = 2000 |
| 8 | 741972 + 3212000 = 3953972 | 2000 − 1000 = 1000 |
| 9 | 3953972 + 3212000 = 7165972 | 1000 − 1000 = 0000 |

We will leave the construction of a flow diagram for this procedure as an exercise.

Besides saving computer-time fewer steps may help increase the accuracy of a solution. Computers do not use the ordinary real numbers because infinite decimal numbers cannot be stored. Instead, computers use a system similar to scientific notation (floating point numbers) with a fixed number of significant digits. Each time an operation is performed, the result is rounded off if necessary to this fixed number of significant digits. The greater the number of operations, the more rounding that occurs, and then there is usually less accuracy.

A good way to show the difference in accuracy is to evaluate a polynomial. Often it is necessary to evaluate an expression like $5x^4 + 3x^3 - 2x^2 + 7x + 3$ for $x$ equal to some particular number. One way to do this is to compute it as it is written using the following steps:

1. Select $x$
2. Multiply $x$ by $x$ and store $x^2$
3. Multiply $x^2$ by $x$ and store $x^3$
4. Multiply $x^3$ by $x$ and store $x^4$
5. Multiply $x$ by 7 and store $7x$
6. Multiply $x^2$ by 2 and store $2x^2$
7. Multiply $x^3$ by 3 and store $3x^3$
8. Multiply $x^4$ by 5 and store $5x^4$
9. Add $5x^4$ and $3x^3$
10. Subtract $2x^2$ from last sum
11. Add $7x$ to last difference
12. Add 3 to last sum

If we were using ordinary computation, with no rounding off, this is perfectly acceptable. However, let's suppose that we are rounding off to two decimal places at each step (a computer actually rounds off to a certain number of significant digits; rounding off to two decimal places will be sufficient to illustrate our point). We will round down if the digit in the third decimal place is less than 5 and will round up if the digit in the third decimal place is 5 or greater.

***Evaluate $5x^4 + 3x^3 - 2x^2 + 7x + 3$ for $x = 2.07$ with no rounding off. Evaluate the expression again for $x = 2.07$ and round off to two decimal places at each step.***

Without rounding, $5x^4 + 3x^3 - 2x^2 + 7x + 3$ is 127.33126905 for $x = 2.07$. With rounding, we get 127.21, which is more than 0.12 from the actual value. We could increase the accuracy if it were possible to evaluate the expression using fewer steps. Such an evaluation is possible because

$$5x^4 + 3x^3 - 2x^2 + 7x + 3 =$$
$$x(5x^3 + 3x^2 - 2x + 7) + 3 =$$
$$x(x[5x^2 + 3x - 2] + 7) + 3 =$$
$$x(x[x\{5x + 3\} - 2] + 7) + 3.$$

This means that we could use the following steps to evaluate the expression:

1. Select $x$
2. Multiply $x$ by 5
3. Add 3
4. Multiply by $x$
5. Subtract 2
6. Multiply by $x$
7. Add 7
8. Multiply by $x$
9. Add 3

***Use this second approach to evaluate the same expression for $x = 2.07$, first without rounding, and then rounding to two decimal places at each step.***

This time, without rounding, we again get 127.33126905, which we would expect since we haven't changed the polynomial. With rounding, however, we get 127.30 which is slightly more than 0.03 from the actual value—a decided increase in accuracy.

Accuracy is one of the major problems in computer work. It is not enough to learn some programming language and then translate some formula from a pure mathematics treatise into this language. The mathematics of a computer is approximate and special approaches must often be developed in order to obtain acceptably accurate results.

## Exercises

1. Find $67 \times 13$ using the repeated addition developed in Section 3 and round off each step to two significant digits. Find $67 \times 13$ using the modified repeated addition with shift described in Section 4, and round off to two significant digits at each step. Compare the accuracy.
2. Construct a flow diagram for multiplication by repeated addition with a shift in position as explained in Section 4.
3. Evaluate $6x^3 - 2x^2 - 4x + 1$ for $x = 2.03$ by the two methods developed in this section and with rounding at each step to two decimal places. Evaluate the expression by either method for $x = 2.03$ without rounding. Which of the two methods is more accurate?
4. For a further discussion of accuracy read the following article: "Solving a Quadratic Equation on a Computer," by George E. Forsythe, pp. 138–152 in *The Mathematical Sciences,* Edited by COSRIMS, M.I.T. Press, Cambridge, Mass., 1969.

## 5. USES OF COMPUTERS

From our discussion so far we would expect to use computers to do long computations. For example, it is not surprising that computers are used for the lengthy computations needed for space flights, for building bridges, for compiling mathematical tables used in applied mathematics, for interpreting the statistical data of an educational psychologist, and for computing interest compounded daily in banks.

Computers also are used on jobs which involve some computations on a large input of information which must be sorted, alphabetized, or tabulated. This is called data processing. Much of the equipment needed to handle punched cards for data processing was developed long before computers. Punched cards were first used for the U.S. census in 1890. Much data processing can be done on office equipment far less complicated than a digital computer. Since information often enters a computer by punched cards, some of this card handling equipment is part of a digital computer. If available it is often convenient to do basic data processing (like payrolls) on a computer and, of course, many data processing jobs (as election predictions) are so complex that a computer is a necessity. Some examples of data processing are census work, charge accounts, college records, inventories, and records of stock transactions.

Besides these uses of a computer, there are many other applications. One somewhat surprising area in which computers are used is pure mathematics. It might be surprising because in pure mathematics we are interested in general cases. It does not seem that the special cases accessible to a computer by computation would be of much value. However, mathematicians often experiment with special cases in order to arrive at a conjecture. With a computer, this work can be done quickly and for many more cases than was possible before. In number theory and combinatorial analysis especially, there are many new theorems which have been stimulated by computer experimentation. Besides inspiring a conjecture, computers can be used to prove that certain conjectures are false. A conjecture can be proved false by producing just one example for which it does not hold. Computers are sometimes able to find this counterexample.

Computers have even been used to prove geometry theorems. They can be programmed to perform the simple logical operations of the two-valued logic of sentences (see Chapter 5). Using these operations, some of the elementary theorems of Euclidean geometry and some theorems of projective geometry have been proved by computers. Even the more difficult theorem of Pappus (see Section 3, Chapter 3) has been proved by a computer. In this case, however, the proof was not a synthetic proof from axioms, but rather an analytic proof. Points and lines are associated with numbers, and theorems are proved by computing with these numbers. An analytic proof of Pappus is an extremely long computation. The interesting thing about this work is not that the computer successfully produced a proof of the theorem, but that in the process the readout inspired some new theorems for which no one had been looking. This influence of the computer is different from that when it assists a mathematician carry out experiments he has designed.

Computers have had another effect on mathematics. They have stimulated the study of many new research problems in pure mathematics. The design of computers, the formulation of programs, the design of algorithms, the questions of solvability of a problem, and problems of accuracy all lead to fundamental mathematical questions.

There have also been many surprising applications of computers to science—applications that are not basically numerical. For example, X-ray diffraction photographs of large molecules are taken to determine the structure of these molecules. The photographs are painstakingly measured for intensity, many calculations are performed on this data, and much trial and error is used to find a three-dimensional model that fits all the data. Almost this entire process can now be done by computers. The photographs are scanned by a photomultiplier tube which measures intensity, this is converted by machine to digital form and punched on cards. The information is then fed into a computer, which does the necessary calculations. A program then takes the outcome of the computations and changes it into a model, which appears on the screen of an oscilloscope.

Scanning by computer is also used to intensify bubble-chamber photographs of elementary particles, to aid in locating brain tumors by radioactive isotopes, and in some chromosome analysis.

Another way computers are used in science is by simulating events. Here, the computer is not just an instrument used in an experiment, but the actor of the experiment itself. Many experiments carried out in simulated form are not even possible to carry out in practice. Launches and orbits of satellites, the impact on a fluid of a projectile travelling at a speed greater than sound, the functioning of a proposed power network, the changes in environment caused by the building of dams, and the flow of customers through a planned supermarket are all simulated by computer.

Composition of music and playing out a boxing match between two men at their primes (even though those primes occurred at different points of history) are other examples of simulation by computer. Pyschologists are extremely hopeful that the use of computers as vicarious behavers is going to lead to new general theories of complex behavior. So far, this type of research has been mostly in the areas of learning, thinking, and decision theory. Computers have been programmed to prove theorems in symbolic logic, recognize patterns, decode Morse code, play checkers and chess, and solve conceptual problems on I.Q. tests. These simulations of behavior have already given new impetus to the study of man's thought processes.

The uses of computers mentioned in this section are only a few of the many which exist. Computers are also used to arrange airline reservations, compute bank balances, retrieve information in all fields of research, control quality in manufacturing, predict the weather, compute TV ratings, assist in programmed learning, and control traffic. Even Babbage would be amazed at what his brainchild has accomplished.

## Exercises

**READ ONE OR MORE OF THE FOLLOWING**

Cerutti, Elsie, and P. J. Davis, "Formac Meets Pappus," *The American Mathematical Monthly,* Vol. 76, No. 8, October, 1969, pp. 895–905.

Churchhouse, R. F., and J.-C. Herz, Editors, *Computers in Mathematical Research,* North-Holland, Amsterdam, 1968. (Most of the articles in this collection are technical and some are in French; however, the first article by D. H. Lehmer on "Machines and Pure Mathematics" is expository. There is also an interesting bibliography at the end of the book; it lists all the published papers on mathematical research done before 1967 with the aid of a computer.)

Coons, Steven Anson, "The Uses of Computers in Technology," *Scientific American,* Vol. 215, No. 3, September, 1966, pp. 176–188.

Fox, L., Editor, *Advances in Programming and Non-Numerical Computation,* Pergamon, Oxford, 1966.

Greenberger, Martin, "The Uses of Computers in Organizations," *Scientific American,* Vol. 215, No. 3, September, 1966, pp. 192–202.

Hamming, R. W., "Impact of Computers," *The American Mathematical Monthly,* Vol. 72, No. 2, Part II, February, 1965, pp. 1–7.

Lipetz, Ben-Ami, "Information Storage and Retrieval," *Scientific American,* Vol. 215, No. 3, September, 1966, pp. 224–242.

*Oettinger, Anthony G., "The Uses of Computers in Science," *Scientific American,* Vol. 215, No. 3, September, 1966, pp. 160–172.

Shannon, Claude E., "A Chess-Playing Machine," in *The World of Mathematics,* Vol. 4, J. R. Newman, Simon and Schuster, New York, 1956, pp. 2124–2133.

Suppes, Patrick, "The Uses of Computers in Education," *Scientific American,* Vol. 215, No. 3, September, 1966, pp. 206–220.

*Ulam, Stanislaw M., "Computers," *Scientific American,* Vol. 211, No. 3, September, 1964, pp. 202–216.

## 6. WHAT IS COMPUTER SCIENCE?

Briefly, as we said before, computer science is the study of computers. It includes the design of *hardware,* the components of computers and the total systems, and the design of *software,* languages, compilers, ways to debug programs (precompilers), and development of useful algorithms. The computer scientist must be concerned with practical questions like speed, economy, accuracy, and the hardware and software required for a particular problem. He is also becoming more concerned with theoretical problems: general principles of problem solving, questions of existence of solutions, information processing theory, and possible future computers.

The subject of possible future computers often stimulates a discussion on the similarities and differences between a computer and the human brain. Computers already have "memories" and their storage capabilities in the future will certainly exceed that of the human brain. However, the access rate of a computer is much slower than that of the brain and it is hard to see that this will be overcome in the future. Computers do have the advantage of not forgetting, not being given to fatigue, and not depending on emotional motivation to undertake a huge task. Computers equipped with optical scanners already can "see;" however, their ability to recognize patterns is very crude compared to the ability of human beings. Computers have been programmed to "learn." For example, checker-playing skills of computers have been increased by studying the games already played. Computers, however, have not been able to discover similarities and learn general principles.

Computers do use languages. At present they are limited to rather short lists of technical expressions. Programming computers to handle natural languages has not progressed very far. Although machines do aid in translating from one natural language to another, a complete translation done by computer still is not feasible.

There has been some almost unbelievable work done on possible future computers. Turing has shown that a universal machine can be built that will imitate any other existing machine. Von Neumann has shown that it is possible to build a machine that can "reproduce" itself.

It is hard to judge now whether such machines will ever be built. However, these studies are not just fascinating speculations. One of the outgrowths already is the field of cybernetics, in which information-handling machines are compared with the nervous system of men and animals. Even if the futuristic machines never get built, the study of their possibility has led to a greater understanding of the human information-processing system.

Can machines think? Are humans merely sophisticated machines? These are important questions that no one has yet adequately answered. We can safely say that, just as the telescope extends the capabilities of our eyes and jet planes extend our power of locomotion, computers have certainly extended human intelligence.

## Exercises

**READ ONE OR MORE OF THE FOLLOWING**

Bernstein, Jeremy, *The Analytical Engine: Computers—Past, Present, and Future,* Random House, New York, 1964, pp. 81–103.

Forsythe, George E., "What To Do till the Computer Scientist Comes," *The American Mathematical Monthly,* Vol. 75, No. 5, May, 1968, pp. 454–462.

*Kemeny, John G., "Man Viewed as a Machine," *Scientific American,* Vol. 192, No. 4, April, 1955, pp. 58–67.

Minsky, Marvin L., "Artificial Intelligence," *Scientific American,* Vol. 215, No. 3, September, 1966, pp. 246–260.

Schwartz, J. T., "Prospects of Computer Science," *The Mathematical Sciences,* Edited by COSRIMS, M.I.T. Press, Cambridge, Mass., 1969, pp. 261–271.

Turing, A. M., "Can a Machine Think?" in *The World of Mathematics,* Vol. 4, J. R. Newman, Simon and Schuster, New York, 1956, pp. 2099–2123.

Von Neumann, John, *The Computer and the Brain,* Yale University Press, New Haven, 1958.

Von Neumann, John, "The General and Logical Theory of Automata," in *The World of Mathematics,* Vol. 4, J. R. Newman, Simon and Schuster, New York, 1956, pp. 2070–2098.

## Test

**1.** In each of the following specify the input, operation, and output:

(a) playing a tape recorder
(b) gumball machine
(c) telegraph
(d) hourglass
(e) counter used to tally attendance
(f) a magnifying glass

**2.** Which of the following devices are of an analog nature, and which digital: an electronic watch, a class attendance chart, a mercury thermometer, a light meter?

**3.** The average of $n$ numbers, $a_1, a_2, \ldots, a_n$ is $(a_1 + a_2 + \ldots + a_n)/n$. Construct a flow diagram for finding the average of any finite set of numbers on a computer whose operations include addition and division. Notice that the program will have to determine what $n$ is for any given set.

**4.** Discuss each of the following statements:

(a) Computers merely do arithmetic.
(b) Computers do only what they are told to do.
(c) Computers can do only those things that we ourselves can do.
(d) Computers will someday duplicate all the capabilities of human intelligence.

**5.** What outside reading did you do about computers or their applications? Summarize anything additional to the text which you learned from this reading.

# CONCLUSION

## What Is Mathematics?

Let's return now to our original question: What is mathematics? We will not be so foolish as to attempt a short summary answer here. This whole book is our attempt at an answer. Our answer could be summarized best by rereading the last two sections of every chapter.

Instead, let's end by giving the question a more personal twist: "What is mathematics to me?" "Now that I've finished this book (or this course), what should I carry away from it?"

***Stop here and try to answer this yourself.***

First of all, hopefully, this close-up view of mathematics has removed any negative attitude we may have had: "I hate mathematics." "I'm afraid to take a mathematics course." "Mathematics and I never did get along." "I think math is one of the most useless subjects I've ever taken." "Mathematics is so dry and unimaginative."

Even if we do not take part in any professional field directly associated with mathematics, we cannot ignore it. Our children will study it in school; hopefully, we are now in a position to help them understand why they need mathematics. We may get to be members of local school boards; the experience of this book should help us function more intelligently. Certainly we will be tax-payers whose money supports mathematical education and research; it may relieve the burden a little to know the value of what the money is supporting.

Our discussions should have given us some insight into what mathematicians do, some appreciation of mathematics as a creative art, and much respect for the role of mathematics in the practical achievements of mankind. Now that we ourselves have had a close look at some mathematics, the title of the book, *Mathematics: Art and Science,* should be meaningful. Mathematics is an art, often pursued for its beauty, harmony, and creativity; and mathematics is a science, constantly being applied to the problems and needs of humanity. With this dual role there can be no doubt that mathematics is an enduring and relevant field of study.

# Hints and Answers to Selected Exercises

## CHAPTER 1

### Section 1

2. $d(26) = 4$ $d(27) = 4$ $d(28) = 6$ $d(29) = 2$ $d(30) = 8$ $d(31) = 2$ $d(32) = 6$ $d(33) = 4$
   $d(34) = 4$ $d(35) = 4$ $d(36) = 9$ $d(37) = 2$ $d(38) = 4$ $d(39) = 4$ $d(40) = 8$ $d(41) = 2$
   $d(42) = 8$ $d(43) = 2$ $d(44) = 6$ $d(45) = 6$ $d(46) = 4$ $d(47) = 2$ $d(48) = 10$ $d(49) = 3$ $d(50) = 6$

4. A statement is false if it is not true in even one specific case.

6. (a) Name a pro-basketball player taller than seven feet.
   (b) $3 + 5 = 8$.
   (c) $1/2 + 1/2 \neq 2/4$.
   (d) A right triangle with angles 90°-45°-45° is not similar to a right triangle with angles 90°-60°-30°.
   (e) $X = 0$.

### Section 2

2. 15, 92, 93, 121, 135.

4. $10 = 3 + 7$ and $10 = 5 + 5$, so the representation of 10 as the sum of two primes is not unique. $18 = 5 + 13$ and $18 = 7 + 11$, so the representation of 18 as the sum of two primes is not unique.

### Section 3

2. $d(16) = 5$ $d(21) = 4$ $d(24) = 8$
   $d(26) = 4$ $d(30) = 8$ $d(32) = 6$
   $d(40) = 8$ $d(44) = 6$ $d(50) = 6$

4. $2^2 - 1 = 3$, $2^3 - 1 = 7$, $2^5 - 1 = 31$, $2^7 - 1 = 127$. This is not a proof because it does not consider every prime $p$. The conjecture is false; $2^{11} - 1 = 2047 = 23 \cdot 89$ is a counterexample.

### Section 4

2. $180 = 2^2 \cdot 3^2 \cdot 5^1$.
   $2^0 \cdot 3^0 \cdot 5^0 = 1$ $2^1 \cdot 3^0 \cdot 5^0 = 2$ $2^2 \cdot 3^0 \cdot 5^0 = 4$

| | | |
|---|---|---|
| $2^0 \cdot 3^0 \cdot 5^1 = 5$ | $2^1 \cdot 3^0 \cdot 5^1 = 10$ | $2^2 \cdot 3^0 \cdot 5^1 = 20$ |
| $2^0 \cdot 3^1 \cdot 5^0 = 3$ | $2^1 \cdot 3^1 \cdot 5^0 = 6$ | $2^2 \cdot 3^1 \cdot 5^0 = 12$ |
| $2^0 \cdot 3^1 \cdot 5^1 = 15$ | $2^1 \cdot 3^1 \cdot 5^1 = 30$ | $2^2 \cdot 3^1 \cdot 5^1 = 60$ |
| $2^0 \cdot 3^2 \cdot 5^0 = 9$ | $2^1 \cdot 3^2 \cdot 5^0 = 18$ | $2^2 \cdot 3^2 \cdot 5^0 = 36$ |
| $2^0 \cdot 3^2 \cdot 5^1 = 45$ | $2^1 \cdot 3^2 \cdot 5^1 = 90$ | $2^2 \cdot 3^2 \cdot 5^1 = 180$ |

Yes, all are divisors of 180. No. 18. Yes, since $d(180) = (2+1)(2+1)(1+1) = 18$.

**4.** 48, 180, 360.

**6.** If $N$ is a perfect square, then all the $a$'s in the formula are even and $d(N)$ is odd. Therefore, if $d(N)$ is even (that is, not odd) then $N$ cannot be a perfect square.

**Section 5**

**1.** 4, 8, 6, 6, 9.

**3.**

| | | | |
|---|---|---|---|
| $s(1) = 1$ | $s(8) = 15$ | $s(14) = 24$ | $s(20) = 42$ |
| $s(2) = 3$ | $s(9) = 13$ | $s(15) = 24$ | $s(21) = 32$ |
| $s(3) = 4$ | $s(10) = 18$ | $s(16) = 31$ | $s(22) = 36$ |
| $s(4) = 7$ | $s(11) = 12$ | $s(17) = 18$ | $s(23) = 24$ |
| $s(5) = 6$ | $s(12) = 28$ | $s(18) = 39$ | $s(24) = 60$ |
| $s(7) = 8$ | $s(13) = 14$ | $s(19) = 20$ | $s(25) = 31$ |

It seems that $s(N)$ is multiplicative.

**5.** Since $a(2) \cdot a(9) = 2 \cdot 9 \neq 9 = a(18)$, $a(N)$ is not multiplicative.

**Section 6**

**2.** $\gcd(1,2) = 1$, $\gcd(2,3) = 1$, $\gcd(4,6) = 2$, $\gcd(8,12) = 4$, $\gcd(8,4) = 4$, $\gcd(3,6) = 3$, $\gcd(4,24) = 4$, and $\gcd(4,4) = 4$. If $A$ and $B$ are related in the lattice by an ascending line, then the $\gcd(A,B)$ is the smaller of $A$ and $B$. If $A \neq B$ and $A$ and $B$ are not connected by an ascending line, then the $\gcd(A,B)$ is the closest number beneath both $A$ and $B$ and attached to each by an ascending line. If $A = B$, then $\gcd(A,B) = A$.

**3.** $\text{lcm}(2,3) = 6$, $\text{lcm}(1,2) = 2$, $\text{lcm}(4,6) = 12$, $\text{lcm}(8,4) = 8$, $\text{lcm}(3,6) = 6$, $\text{lcm}(4,24) = 24$, and $\text{lcm}(4,4) = 4$. If $A$ and $B$ are connected in the lattice by an ascending line, then the $\text{lcm}(A,B)$ is the larger of $A$ and $B$. If $A \neq B$ and $A$ and $B$ are not connected by an ascending line, then the $\text{lcm}(A,B)$ is the closest number above both $A$ and $B$ and attached to each by a line. If $A = B$, then $\text{lcm}(A,B) = A$.

**Section 7**

**1.** (a) $2|6$, but $6\nmid 2$.
(b) $6|4 \cdot 3$ and $6\nmid 4$, but $6\nmid 3$.
(c) $6\nmid 4$ and $6\nmid 3$, but $6|12$.

**3.** Assume that brown is purple and red. Since purple is blue and red, it follows that brown is blue and red and red, or simpler brown is blue and red. Also brown is green and red, and green is yellow and blue, so brown is yellow and blue and red. Both of these descriptions of brown cannot be true since in each case we have reduced brown to primary colors. Therefore, our initial assumption was wrong and brown is not purple and red.

## CHAPTER 2

### Section 1

**1.** $A = \begin{pmatrix} a & b \\ a & b \end{pmatrix}$ and $B = \begin{pmatrix} a & b \\ b & a \end{pmatrix}$. The table is given in Section 2, Chapter 2.

**2.** $A = \begin{pmatrix} a & b & c & d \\ a & b & c & d \end{pmatrix}$ $B = \begin{pmatrix} a & b & c & d \\ b & c & a & d \end{pmatrix}$ $C = \begin{pmatrix} a & b & c & d \\ c & a & b & d \end{pmatrix}$ $D = \begin{pmatrix} a & b & c & d \\ b & d & c & a \end{pmatrix}$

$E = \begin{pmatrix} a & b & c & d \\ d & a & c & b \end{pmatrix}$ $F = \begin{pmatrix} a & b & c & d \\ a & c & d & b \end{pmatrix}$ $G = \begin{pmatrix} a & b & c & d \\ a & d & b & c \end{pmatrix}$ $H = \begin{pmatrix} a & b & c & d \\ c & b & d & a \end{pmatrix}$

$I = \begin{pmatrix} a & b & c & d \\ d & b & a & c \end{pmatrix}$ $J = \begin{pmatrix} a & b & c & d \\ c & d & a & b \end{pmatrix}$ $K = \begin{pmatrix} a & b & c & d \\ b & a & d & c \end{pmatrix}$ $L = \begin{pmatrix} a & b & c & d \\ d & c & b & a \end{pmatrix}$

$M = \begin{pmatrix} a & b & c & d \\ b & a & c & d \end{pmatrix}$ $N = \begin{pmatrix} a & b & c & d \\ c & b & a & d \end{pmatrix}$ $O = \begin{pmatrix} a & b & c & d \\ d & b & c & a \end{pmatrix}$ $P = \begin{pmatrix} a & b & c & d \\ a & c & b & d \end{pmatrix}$

$Q = \begin{pmatrix} a & b & c & d \\ a & d & c & b \end{pmatrix}$ $R = \begin{pmatrix} a & b & c & d \\ a & b & d & c \end{pmatrix}$ $S = \begin{pmatrix} a & b & c & d \\ b & c & d & a \end{pmatrix}$ $T = \begin{pmatrix} a & b & c & d \\ d & a & b & c \end{pmatrix}$

$U = \begin{pmatrix} a & b & c & d \\ b & d & a & c \end{pmatrix}$ $V = \begin{pmatrix} a & b & c & d \\ c & a & d & b \end{pmatrix}$ $W = \begin{pmatrix} a & b & c & d \\ c & d & b & a \end{pmatrix}$ $X = \begin{pmatrix} a & b & c & d \\ d & c & a & b \end{pmatrix}$

**3.** The table is given in Section 2, Chapter 2.

### Section 2

**1.** (a) $B$ does not appear in each row and column; no row and no column looks like the guide row and guide column.

(b) No row and no column looks like the guide row and guide column; the operation is not associative because $A(BC) \neq (AB)C$.

(c) $D$ and $E$ are not in the guide row and column.

(d) No column looks like the guide column; the operation is not associative because $C(CC) \neq (CC)C$.

(e) No row looks like the guide row; the operation is not associative because $A(BC) \neq (AB)C$.

(f) The operation is not associative because $(BB)C \neq B(BC)$.

**3.** If the rearrangements are symbolized as follows:

$A = \begin{pmatrix} a & b & c & d \\ a & b & c & d \end{pmatrix}$ $S = \begin{pmatrix} a & b & c & d \\ b & c & d & a \end{pmatrix}$ $J = \begin{pmatrix} a & b & c & d \\ c & d & a & b \end{pmatrix}$ $T = \begin{pmatrix} a & b & c & d \\ d & a & b & c \end{pmatrix}$

then the table is

| | A | S | J | T |
|---|---|---|---|---|
| A | A | S | J | T |
| S | S | J | T | A |
| J | J | T | A | S |
| T | T | A | S | J |

These products are the same as those given in the table for Group 3.

**5.**

| | 0 | 1 | 2 | 3 | 4 | 5 | 6 |
|---|---|---|---|---|---|---|---|
| 0 | 0 | 1 | 2 | 3 | 4 | 5 | 6 |
| 1 | 1 | 2 | 3 | 4 | 5 | 6 | 0 |
| 2 | 2 | 3 | 4 | 5 | 6 | 0 | 1 |
| 3 | 3 | 4 | 5 | 6 | 0 | 1 | 2 |
| 4 | 4 | 5 | 6 | 0 | 1 | 2 | 3 |
| 5 | 5 | 6 | 0 | 1 | 2 | 3 | 4 |
| 6 | 6 | 0 | 1 | 2 | 3 | 4 | 5 |

This table does specify a group. If the days of the week beginning with Sunday are numbered 0, 1, 2, 3, 4, 5, 6, then five days after Thursday can be found by considering $4+5=2$, that is, Tuesday.

**7.** (a) The identity is 0. For all $x$, $x' = -x$.
(b) The identity is 0. For all $x$, $x' = -x$.
(c) The identity is 1. For all $x$, $x' = 1/x$. 0 has no inverse.
(d) The identity is 0. For all $x$, $x' = -x$.
(e) The identity is 1. For all $x$, $x' = 1/x$.

### Section 3

**1.** $\{A,L\}$, $\{A,M\}$, $\{A,N\}$, $\{A,O\}$, $\{A,P\}$, $\{A,Q\}$, and $\{A,R\}$.

**3.** The table is given in Section 4, Chapter 2. Properties 1 to 6 are apparent in this table. Property 7 does not have to be checked since associativity is inherited from Group 3.

**5.** No. Groups 2 and 3 are not commutative. In Group 2, $BD = E \neq F = DB$. In Group 3, $BD = L \neq J = DB$.

### Section 4

**2.** $A\{A,S,J,T\} = \{A,S,J,T\}$, $S\{A,S,J,T\} = \{S,J,T,A\}$, $J\{A,S,J,T\} = \{J,T,A,S\}$, $T\{A,S,J,T\} = \{T,A,S,J\}$, $N\{A,S,J,T\} = \{N,L,Q,K\}$, $Q\{A,S,J,T\} = \{Q,K,N,L\}$, $K\{A,S,J,T\} = \{K,N,L,Q\}$ and $L\{A,S,J,T\} = \{L,Q,K,N\}$.

**4.** The union of $\{A,L\}$, $\{S,N\}$, $\{J,K\}$, and $\{T,Q\}$.

**6.** Four. $8/2 = 4$.

### Section 5

**1.** Let $\mathbf{H} = \{A,K\}$ and $\mathbf{G} = \{A,S,J,T,N,Q,K,L\}$ with the table of Section 4. **H** contains two elements. If $A_i\mathbf{H}$ is a left coset of **H** relative to **G**, $A_i\mathbf{H} = \{A_iA, A_iK\}$; for example, if $A_i = S$, $S\mathbf{H} = \{SA, SK\} = \{S, Q\}$. This coset is a part of the $S$ row of the table for **G**, $S,J,T,A,K,L,Q,N$. Every element of **G** appears only once in this row, so the two elements in $S\mathbf{H}$ are distinct. This means that $S\mathbf{H}$ has exactly two elements.

**3.** Yes. They could contain 1, 2, 5, or 10 elements because all of these numbers divide 10 and the Lagrange Theorem would be satisfied. No, the Lagrange Theorem says that *if* there is a subgroup the number of elements in that subgroup divides the number of elements in the group. The Lagrange Theorem does not imply that there is a subgroup for every divisor.

**5.** If $XS = S$ for all elements $S$ of the group, then the $X$ row of the table looks like the guide row. From the last exercise, the $X$ column of the table looks like the guide column; therefore, $SX = S$ for all elements $S$ of the group.

## Section 6

**2.** *Conjecture:* If a subgroup **H** of a group **G** has half as many elements in it as the group, then this subgroup is normal.
*Hint:* Show that there are only two distinct left cosets, **H** and the set of elements in **G** which are not in **H**. Any element $A_i$ in **H** gives the left coset $A_i\mathbf{H} = \mathbf{H}$. Any element $A_k$ not in **H** gives the other left coset. Show that a similar thing happens for the right cosets. Finally, show that for any $A$ in **G**, $A\mathbf{H} = \mathbf{H}A$.

**4.** $\{A\}$, $\{A,D\}$, $\{A,C,E\}$, and $\{A,B,C,D,E,F\}$. Form all left and all right cosets of each subgroup and compare the corresponding cosets.

**6.** $\{A,J\}$. $\{A,J\}$, $\{A,J,K,L\}$. Let subgroups **H** and **K** be considered. If **H** and **K** are related by an ascending line, then the largest common subgroup of **H** and **K** is the smaller of **H** and **K**. If **H** and **K** are not connected by ascending lines and $\mathbf{H} \neq \mathbf{K}$, then the largest common subgroup of **H** and **K** is the closest subgroup beneath both **H** and **K** and attached to them by ascending lines. If $\mathbf{H} = \mathbf{K}$, then the largest common subgroup is **H**.

**8.** $\mathbf{HK} = \{AA, AN, KA, KN\} = \{A, N, K, S\}$. This is not a subgroup of Group 4. If a lattice was formed from all subgroups instead of only the normal subgroups, we could not use the lattice to find the "product" of two subgroups, because the "product" of two subgroups which are not normal is not a subgroup.

## Section 7

**1.** The two symmetries form a group. The table is like Group 1 of Section 2 with $I = A$ and $R = B$.

**2.** The six rotations are:

$R_0$ = no movement,
$R_1$ = clockwise rotation of 60°,
$R_2$ = clockwise rotation of 120°,
$R_3$ = clockwise rotation of 180°,
$R_4$ = clockwise rotation of 240°,
$R_5$ = clockwise rotation of 300°.

The six reflections are:

$F_1$ = a reflection in the vertical,
$F_2$ = a reflection in the 2 and 8 o'clock line,
$F_3$ = a reflection in the 4 and 10 o'clock line,
$F_4$ = a reflection in the 1 and 7 o'clock line,
$F_5$ = a reflection in the horizontal,
$F_6$ = a reflection in the 5 and 11 o'clock line.

The table of these twelve symmetries is a group table.

## CHAPTER 3

### Section 1

**2.** Quadrilateral.

**4.** Quadrilateral.

**6.** Assume that the second construction leads to a point $P''$. Show that triangle $AOP$ is similar to triangle $BP'P$. Show that triangle $COP$ is similar to triangle $DP''P$. Finally $P'P/OP = P''P/OP$, so $P'P = P''P$ and $P' = P''$.

### Section 2

**3.** Drawings represent special cases; a proof must cover all cases. Drawings help us to understand the situation and lead to conjectures.

**5.** They are concurrent. If two triangles are such that their corresponding sides extended intersect in collinear points, then the lines joining their corresponding vertices are concurrent. This is the converse of Desargues' Theorem; it also looks like Desargues' Theorem with point and line, vertex and side, intersection and line joining, and concurrent and collinear interchanged.

**7.** Yes. Since $AA'$, $BB'$, and $CC'$ are mutually parallel in the Euclidean plane, they are all concurrent at a point at infinity in the projective plane.

### Section 3

**3.** $AD$, $BE$, and $CF$ are concurrent. If a hexagon is such that its alternate sides are concurrent at two points, then the lines joining the opposite vertices are concurrent. This looks like Pappus' Theorem with side and vertex, concurrent and collinear, point and line, line joining and point of intersection, and vertex and side interchanged.

### Section 4

**1.** Use A3.

**3.** Let the given point be $P$ and $A, B, C, D$ be the points in A3. Consider the cases in which $P$ is on three, two, one, and none of the six lines determined by $A$, $B$, $C$, and $D$.

**5.** Some examples are:
Hexagon $ACEBDF$ with $A, E, D$ and $C, B, F$ as alternate vertices and $AC$ and $BD$, $CE$ and $DF$, and $EB$ and $FA$ as opposite sides; Hexagon $ACBDEF$ with $A, B, E$ and $C, D, F$ as alternate vertices and $AC$ and $DE$, $CB$ and $EF$, and $BD$ and $FA$ as opposite sides; Hexagon $ABCDFE$ with $A, C, F$ and $B, D, E$ as alternate vertices and $AB$ and $DF$, $BC$ and $FE$, and $CD$ and $EA$ as opposite sides. (There are other examples.)

### Section 5

**2.** Use triangles $ABC$ and $DEF$.

### Section 6

1. A3 is not true for this model.
5. See Section 2, Chapter 4.
7. If two objects $A$ and $B$ are connected by an ascending line, then the intersection of $A$ and $B$ is the smaller of $A$ and $B$; if $A \neq B$ and they are not connected by an ascending line, then the intersection of $A$ and $B$ is the closest object below both of them and connected to them by ascending lines; if $A = B$, then the intersection is $A$.

## CHAPTER 4

### Section 1

1. (b), (c), (e), (h), and (i) are true.
5. **H, H, K, M; M, L, M, M.**

### Section 2

1.

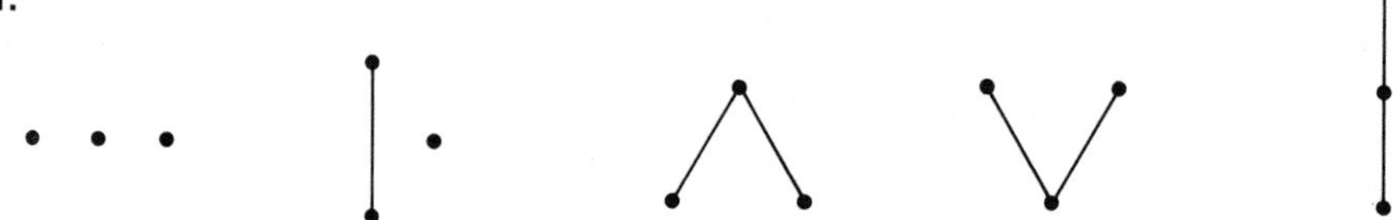

3.

5.

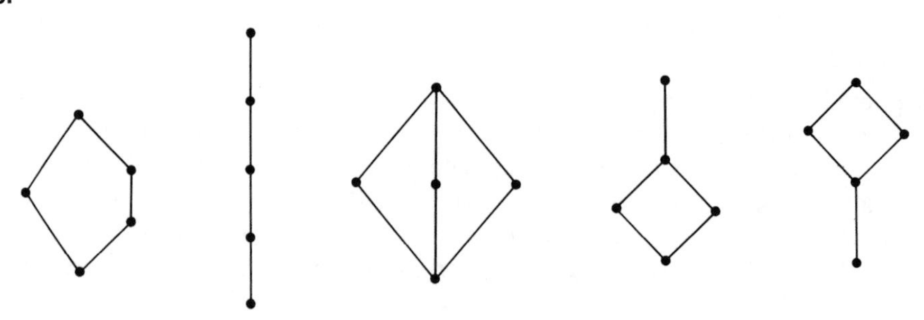

7. $Q$ or $P$; only $P$; $R$, $S$, $P$, or $V$; only $P$.
   $Q$; $P$; $V$; $P$.

### Section 3

1. Let $A$ cup $B = C$ and $A$ cup $B = E$. Then $A \preceq C$, $B \preceq C$, $A \preceq E$, $B \preceq E$, $C \preceq E$, and $E \preceq C$ by Definition 1. If $C \neq E$, $C \preceq E$ and $E \preceq C$ contradicts L1, so $C \neq E$ is false; therefore, $C = E$.
3. The proof is similar to $A$ cup ($B$ cup $C$) = ($A$ cup $B$) cup $C$.
5. If $A$ cup $B = C$, then $A \preceq C$ and $B \preceq C$ and if $A \preceq X$, $B \preceq X$, then $C \preceq X$. This statement means the same as, $B \preceq C$ and $A \preceq C$ and if $B \preceq X$, $A \preceq X$, then $C \preceq X$, so $B$ cup $A = C$. The proof for cap would be similar reversing the objects about $\preceq$.
7. The second property gives $X$ cap ($X$ cup $Y$) $= X$ where $X = C$ and $Y = D$, so $X$ cup $X = X$ cup ($X$ cap ($X$ cup $Y$)). Then using the first property $X$ cup ($X$ cap ($X$ cup $Y$)) $= X$ where $X = A$ and $X$ cup $Y = B$.

### Section 4

1. $A$ cup $B$ is the larger of $A$ and $B$. $A$ cap $B$ is the smaller of $A$ and $B$.
3. **H** cap **K** is the intersection of **H** and **K**.
5. **A** cap **B** is **A** $\cap$ **B**.

## CHAPTER 5

### Section 1

1. The reasoning seems valid. Since snow is white, the conclusion is not true.
   Let $p$ stand for "Feathers are soft."
   Let $q$ stand for "Snow is blue."
   Let $r$ stand for "Roses are green."
   The argument then has the form:
   PREMISES
   1. $p$.
   2. If not-$q$, then $r$.
   3. Not both $p$ and $r$.

   CONCLUSION: $q$.
   This argument is similar to the examples in Section 1 except that the roles of $q$ and not-$q$ have been exchanged.

### Section 2

1. Tom and Sue, Bob and Mary.
3. Ken owns a Chevy, Bill a Ford, and Jim an Olds.

### Section 3

1. (c), (d), (e), (f), and (h).
3. A1 with $X \rightarrow (X \vee Y) = A$ and $X \wedge Y = B$;
   A6 with $X = A$ and $Y = B$;
   D applied to 1 and 2;

A2 with $X \wedge Y = A$, $X = B$, and $X \vee Y = C$;
D applied to 3 and 4;
A4 with $X = A$ and $Y = B$;
D applied to 5 and 6.

**5.** 1. $[X \to (X \vee Y)] \to \{(Y \wedge X) \to [X \to (X \vee Y)]\}$ from A1 with $X \to X \vee Y = A$ and $Y \wedge X = B$.
2. $X \to (X \vee Y)$ from A6 with $X = A$ and $Y = B$.
3. $(Y \wedge X) \to [X \to (X \vee Y)]$ from 1, 2, and D.
4. $\{(Y \wedge X) \to [X \to (X \vee Y)]\} \to \{[(Y \wedge X) \to X] \to [(Y \wedge X) \to (X \vee Y)]\}$ from A2 with $Y \wedge X = A$, $X = B$, and $X \vee Y = C$.
5. $[(Y \wedge X) \to X] \to [(Y \wedge X) \to (X \vee Y)]$ from 3, 4, and D.
6. $Y \wedge X \to X$ from A5 with $Y = A$ and $X = B$.
7. $(Y \wedge X) \to (X \vee Y)$ from 5, 6, and D.

## Section 4

**3.** (c), (d), (f), (g), (k), (l), and (m).

## Section 5

**2.**

| $A$ | $\neg A$ |
|---|---|
| 1 | 0 |
| 2/3 | 1/3 |
| 1/3 | 2/3 |
| 0 | 1 |

| $\to$ | 1 | 2/3 | 1/3 | 0 |
|---|---|---|---|---|
| 1 | 1 | 2/3 | 1/3 | 0 |
| 2/3 | 1 | 1 | 2/3 | 1/3 |
| 1/3 | 1 | 1 | 1 | 2/3 |
| 0 | 1 | 1 | 1 | 1 |

| $\vee$ | 1 | 2/3 | 1/3 | 0 |
|---|---|---|---|---|
| 1 | 1 | 1 | 1 | 1 |
| 2/3 | 1 | 2/3 | 2/3 | 2/3 |
| 1/3 | 1 | 2/3 | 1/3 | 1/3 |
| 0 | 1 | 2/3 | 1/3 | 0 |

| $\wedge$ | 1 | 2/3 | 1/3 | 0 |
|---|---|---|---|---|
| 1 | 1 | 2/3 | 1/3 | 0 |
| 2/3 | 2/3 | 2/3 | 1/3 | 0 |
| 1/3 | 1/3 | 1/3 | 1/3 | 0 |
| 0 | 0 | 0 | 0 | 0 |

# CHAPTER 6

## Section 1

**1.** (a) { }
(b) { }, {1}.
(c) { }, {1}, {2}, {1,2}.
(d) { }, {1}, {2}, {3}, {1,2}, {1,3}, {2,3}, {1,2,3}.
(e) { }, {1}, {2}, {3}, {4}, {1,2}, {1,3}, {1,4}, {2,3}, {2,4}, {3,4}, {1,2,3}, {1,2,4}, {1,3,4}, {2,3,4}, {1,2,3,4}.
(f) { }, {1}, {2}, {3}, {4}, {5}, {1,2}, {1,3}, {1,4}, {1,5}, {2,3}, {2,4}, {2,5}, {3,4}, {3,5}, {4,5}, {1,2,3}, {1,2,4}, {1,2,5}, {1,3,4}, {1,3,5}, {1,4,5}, {2,3,4}, {2,3,5}, {2,4,5}, {3,4,5}, {1,2,3,4}, {1,2,3,5}, {1,2,4,5}, {1,3,4,5}, {2,3,4,5}, {1,2,3,4,5}.

**3.** As subsets are formed the twenty-six letters must be considered and for each a decision must be made to use it or not use it as a member of the subset. For example, if the subset $\{a,c\}$ is formed, the letter $a$ is used, $b$ is not used, $c$ is used, and the remaining twenty-three letters are not used. Since there are two possibilities for each letter (used, or not used) and there are twenty-six independent decisions, the total number of possible combinations is $2^{26}$, that is, $2^{26}$ different subsets.

**Section 2**

**1.** $\#(\mathbf{A}) = \#(\mathbf{M})$ because of the following 1-to-1 correspondence: $1 \leftrightarrow 3$, $2 \leftrightarrow 4$, $3 \leftrightarrow 5$, $4 \leftrightarrow 6, \ldots, n \leftrightarrow n+2, \ldots$.

**3.** $\#(\mathbf{A}) = \#(\mathbf{O})$ because $\mathbf{O} = \{0,1,-1,2,-2,\ldots\}$ and the following 1-to-1 correspondence exists: $1 \leftrightarrow 0$, $2 \leftrightarrow 1$, $3 \leftrightarrow -1$, $4 \leftrightarrow 2$, $5 \leftrightarrow -2, \ldots, n \leftrightarrow n/2$ if $n$ is even and $n \leftrightarrow (1-n)/2$ if $n$ is odd.

**5.** $\#(\mathbf{A}) = \#(\mathbf{Q})$ because of the 1-to-1 correspondence $n \leftrightarrow -n$.

**7.** $\#(\mathbf{A}) = \#(\mathbf{S})$ because of the 1-to-1 correspondence $n \leftrightarrow n^3$.

**Section 3**

**2.** Let $\mathbf{Y} = \{\{1\}, \{2\}, \{3\},\ldots\}$, $\mathbf{Y}$ is in $\mathscr{P}(\mathbf{X})$ where $\mathbf{X} = \{1,2,3,\ldots\}$. The 1-to-1 correspondence is $\{1\} \leftrightarrow 1$, $\{2\} \leftrightarrow 2$, $\{3\} \leftrightarrow 3, \ldots, \{n\} \leftrightarrow n, \ldots$.

**4.** $\mathbf{Y}_t = \{\ \}$ and $\{\ \} \subseteq \mathbf{X}$, so $\mathbf{Y}_t$ is in $\mathscr{P}(\mathbf{X})$.

**6.** $\{\ \}$, $\{4,6,11\}$, $\{2,4,6,8,\ldots\}$;
$\{\{\ \}\}$, $\{\{\ \}, \{1\}, \{2,5,6,10\}\}$, $\{\{1\}, \{2\}, \{3\}, \ldots\}$;
$\{\{\{\ \}\}\}$, $\{\{\{1\}, \{2\}\}, \{\{3,4\}, \{5,6,7\}\}\}$,
$\{\{\{1\}, \{2\}, \{3\}, \ldots\}, \{\{1,2\}, \{3,4\}, \ldots\}, \{\{1,2,3\}, \{4,5,6\}, \ldots\}\}$.

**Section 4**

**1.** 1, 10, 11, 100, 101, 110, 111, 1000, 1001, 1010, 1011, 1100, 1101, 1110, 1111, 10000, 10001, 10010, 10011, 10100, 10101, 10110, 10111, 11000, 11001.

**3.** 15, 42, 4, 101, 9, 23, 152, 102.

**5.**

(a)

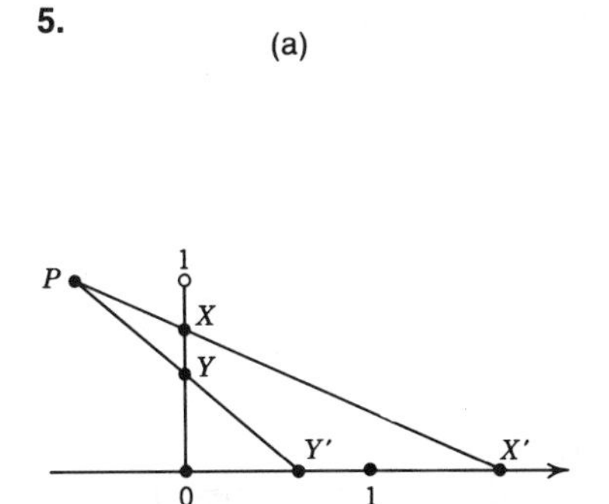


(b)

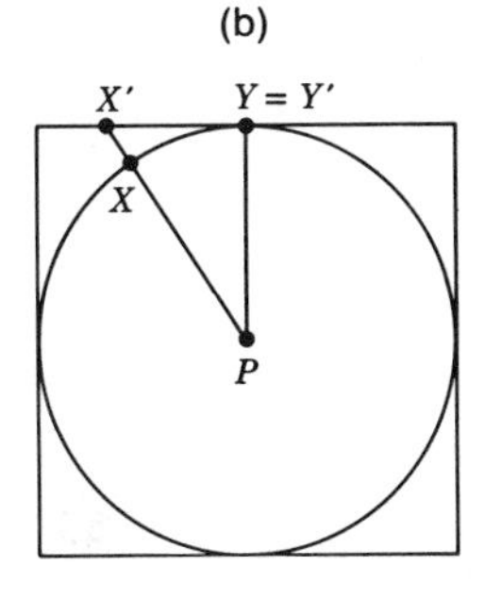


(c)

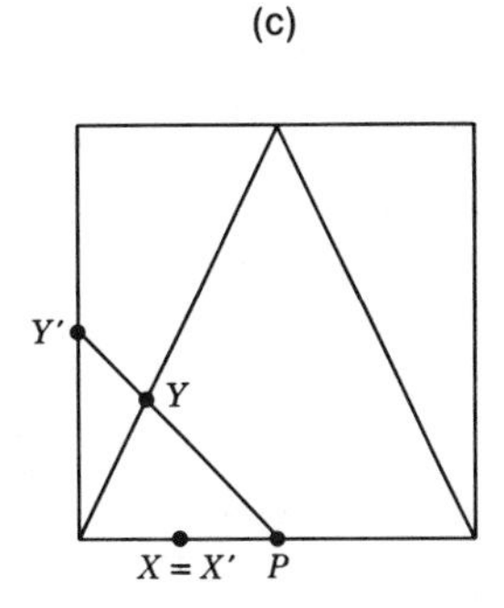

**Section 5**

**1.** (a) $\{1,2,3,4,5\} = 5$

(b) $\{1,2,..,32\} = 32$

(c) $\{1,2,...,50\} = 50$

(d) $\{1,2,3,...\} = \aleph_0$

(e) $\{1,2,3,...\} = \aleph_0$

(f) $\mathscr{P}(\{1,2,3,...\}) = \aleph_1$

**3.** (a) 3, 2, 5. $\#(\mathbf{X}) + \#(\mathbf{Y}) = \#(\mathbf{X} \cup \mathbf{Y})$.

(b) 3, 2, 4. If **X** and **Y** have no elements in common, then $\#(\mathbf{X}) + \#(\mathbf{Y}) = \#(\mathbf{X} \cup \mathbf{Y})$.

(c) 1, $\aleph_0$, $\aleph_0$. $\aleph_0$.

(d) $\aleph_0$, $\aleph_0$, $\aleph_0$. $\aleph_0$.

**CHAPTER 7**

**Section 1**

**1.** (a) Every odd-numbered term is $-2$; every even-numbered term is 2. Six terms.
(c) Every term is 3. Five terms.
(e) The terms are the successive positive multiples of 3. Five terms.

**2.** (a) $[(-2)+2]+[(-2)+2]+[(-2)+2]=0+0+0=0$.
(c) $5 \cdot 3 = 15$.
(e) $(3+6+9+12+15)+(3+6+9+12+15)=(3+15)+(6+12)+(9+9)+(12+6)+(15+3)=5 \cdot 18 = 90$. Therefore, the given series has sum $90/2 = 45$.

**3.** (a) $(-1)^N 2$. (c) 3. (e) $3N$.

**4.** (a) If $N$ is even, the sum is 0; if $N$ is odd, the sum is $-2$.
(c) $3N$.
(e) Let $S = 3+6+9+\ldots+3N$, then

$$2S = (3+3N)+[6+(3N-3)]+\ldots+[(3N-3)+6]+(3N-3)$$
$$= N(3+3N) = 3N+3N^2. \text{ Therefore, } S = 3(N+N^2)/2.$$

**6.** $1/(1 \cdot 2)+1/(2 \cdot 3)+1/(3 \cdot 4)+1/(4 \cdot 5)+1/(5 \cdot 6)=(1/1-1/2)+(1/2-1/3)+(1/3-1/4)+(1/4-1/5)+(1/5-1/6)=1-1/6=5/6$. In a similar way the series with $N$ terms has sum $1 - 1/(N+1) = N/(N+1)$. If $N = 45$, the sum is 45/46.

**8.** $1(1) + 2(1 \cdot 2) + 3(1 \cdot 2 \cdot 3) + \ldots + N(1 \cdot 2 \cdot 3 \ldots N) = [(1 \cdot 2) - (1)] + [(1 \cdot 2 \cdot 3) - (1 \cdot 2)] + \ldots + \{[1 \cdot 2 \cdot 3 \ldots N(N+1)] - (1 \cdot 2 \cdot 3 \ldots N)\} = [1 \cdot 2 \cdot 3 \ldots N(N+1)] - 1$. If $N = 7$, then the sum is 40319.

**Section 2**

**1.** (a) The first term is 2/4; each succeeding term is 1/4 more than the preceding one.
(c) Every term is 1.
(e) The first term is 1, all other terms are 1/3 of the preceding one.
(g) The numerators are the successive positive odd numbers, the denominators are the successive non-negative powers of 2.

**2.** (a) 6/4.
(c) 1.
(e) 1/81.
(g) 9/16.

**3.** (a) $(N+1)/4$.
(c) 1.
(e) $1/3^{N-1}$.
(g) $[1 + (N-1)2]/2^{N-1} = (2N-1)/2^{N-1}$.

**4.** (a) 2/4, 5/4, 9/4, 14/4, 20/4, 27/4, 35/4, 44/4.
(c) 1, 2, 3, 4, 5, 6, 7, 8.
(e) 1, 4/3, 13/9, 40/27, 121/81, 364/243, 1093/729, 3280/2187.
(g) 1, 5/2, 15/4, 37/8, 83/16, 177/32, 367/64, 749/128.

**5.** (a) $N(N+3)/8$ (Use the method of writing the series twice and rearranging the terms.)
(c) $N$.
(e) Let $s_N = 1 + 1/3 + 1/9 + \ldots + 1/3^{N-1}$. Then
$(1/3)s_N = 1/3 + 1/9 + 1/27 + \ldots + 1/3^N$ and
$$s_N - (1/3)s_N = 1 - 1/3^N$$
$$(2/3)s_N = 1 - 1/3^N$$
$$s_N = 3(1 - 1/3^N)/2.$$

**6.** (a) The partial sums increase without bound since every term after the second is greater than or equal to 1.
(c) Increase without bound.
(e) Approach 3/2.
(g) Approach 6.

**Section 3**

**1.** The partial sums are:
(a) 2, 2, 4, 4, 6, 6, 8, 8.

(c) 2/4, 5/4, 9/4, 14/4, 20/4, 27/4, 35/4, 44/4.

(e) −3, 0, −3, 0, −3, 0, −3, 0.

**3.** (a) $1/10^N$.

(b) Let $s_N = 1/10 + 1/10^2 \ldots + 1/10^N$. Then,
$(1/10)s_N = 1/10^2 + 1/10^3 + \ldots + 1/10^{N+1}$.
$s_N - (1/10)s_N = (9/10)s_N = 1/10 - 1/10^{N+1}$. Therefore,
$s_N = 1/9 - 1/(9 \cdot 10^N)$.

(c) Given any $1/10^K$ let $M = K$. Then $1/9 - s_M = 1/9 - [1/9 - 1/(9 \cdot 10^M)] = 1/9 - [1/9 - 1/(9 \cdot 10^K)] = 1/(9 \cdot 10^K) < 1/10^K$.
For any $N \geq M$, $1/9 - s_N < 1/10^K$ since $N \geq K$.

(d) 0.111 . . . .

**5.** For any $1/10^K$ let $M = 4$, $S = 10$, then $S - s_M = 10 - 10 = 0 < 1/10^K$ and for any $N \geq M$, $s_N = 10$, so the same thing is true.

**7.** Let $s_N = 1/2 + 1/4 + 1/8 + \ldots + 1/2^N$. Then
$(1/2)s_N = 1/4 + 1/8 + 1/16 + \ldots + 1/2^{N+1}$
$s_N - (1/2)s_N = (1/2)s_N = 1/2 - 1/2^{N+1}$
$s_N = 1 - 1/2^N$
For any $1/10^K$ let $M = 4K$, then $1 - s_M = 1 - (1 - 1/2^{4K}) = 1/2^{4K} = 1/(2^4)^K < 1/10^K$.
For any $N \geq M$, $N \geq 4K$, so $1 - s_N < 1/10^K$.

**9.** For any $1/10^K$ let $M = 3$, then $1/8 - s_M = 1/8 - (1/10 + 2/10^2 + 5/10^3) = 1/8 - 1/8 = 0 < 1/10^K$. For any $N \geq M$, $s_N = 1/8$, so $1/8 - s_N = 0 < 1/10^K$.

## Section 4

**1.** (a) For any $T$, $s_N > T$ for all $N \geq T + 1$ since each term is 1 or greater.

(c) For any $U > 0$, $s_N < -U$ for all $N \geq U + 1$ because each term is −1.

(e) Since the partial sums oscillate between − 3 and 0 there is no number $S$ which is the "sum" of the series because a small portion of the number line of length $1/10^K$ on either side of $S$ will always exclude an infinite number of partial sums.

(g) The partial sums are −1/8, 7/8, −1/16, 15/16, −1/32, 31/32, −1/64, . . . . The odd partial sums are $s_N = -1/2^{[(N+1)/2+2]}$. The even partial sums are $s_N = [2^{(N/2+2)} - 1]/2^{(N/2+2)}$. The odd partial sums approach 0; the even partial sums approach 1. There are small portions of the number line about any number which will exclude an infinity of these partial sums.

**7.** (a) All the partial sums of $[1 + (-1)] + [1 + (-1)] + \ldots$ are 0, so it converges to 0.

(b) $(1 + 1 + 1) + (1 + 1 + 1) + \ldots$ has partial sums $s_N = 3N$. It does not converge since $s_N$ grows without bound as $N$ does.

(c) $0 + (0 + 0) + (0 + 0 + 0) + \ldots$ has 0 for all its partial sums and so converges to 0.

(d) $[1 + (-3/2)] + [3/4 + (-3/8)] + [3/16 + (-3/32)] + \ldots$ has partial sums −1/2, −1/8, −1/32, −1/128, . . . , that is, $s_N = 1/2^{2N-1}$. This series converges to 0, as does the original series.

(e) From parts (a) and (b) we conjecture that if a series does not converge, we cannot predict what will happen if parentheses are added. From parts (c) and (d), we conjecture that if a series converges, parentheses may be added and it will still converge.

### Section 5

**2.** The series is $9/10 + 9/10^2 + 9/10^3 + \ldots$ with partial sums $9/10$, $99/10^2$, $999/10^3$, $\ldots$ and averages $9/10$, $189/200$, $2889/3000$, $\ldots$. This is $(10 - 1)/10$, $(200 - 11)200$, $(3000 - 111)/3000, \ldots$ so that $t_N = [N \cdot 10^N - (10^0 + 10^1 + 10^2 + \ldots + 10^{N-1})]/(N \cdot 10^N) = 1 - 10^0/(N \cdot 10^N) - 10^1/(N \cdot 10^N) - \ldots - 10^{N-1}/(N \cdot 10^N) = 1 - 1/(N \cdot 10^N) - 1/(N \cdot 10^{N-1}) - \ldots - 1/(N \cdot 10)$. Then $1 - t_N = 1/(N \cdot 10^N) + 1/(N \cdot 10^{N-1}) + \ldots + 1/(N \cdot 10)$. For any $1/10^K$ let $M = 10^K$, then $1 - t_M = (1/M)(0.111 \ldots 1) = (1/10^K)(0.11 \ldots 1)$. For any $N \geq M$, $1 - t_N = (1/N)(0.111 \ldots 1) \leq (1/10^K)(0.11 \ldots 1) < 1/10^K$ because the decimal factor is less than 1.

**4.** The partial sums are $1, -2, 3, -4, 5, -6, \ldots$. In general $s_N = (-1)^{N+1}N$. This does not converge to a "sum" since the odd partial sums are increasing without bound and the even partial sums are decreasing without bound. The averages $1/1, -1/2, 2/3, -2/4, -3/6, 4/7, \ldots$. In general if $N$ is odd, $t_N = (N+1)/2N = 1/2 + 1/2N$, and if $N$ is even $t_N = -N/2N = -1/2$. The original series does not converge according to the first arithmetic mean because the odd partial sums are approaching $1/2$ and the even partial sums always equal $-1/2$.

## CHAPTER 8

### Section 1

**1.** (a) A collection of seeds is planted. Each seed leads to a plant of its own type.
(c) A large number of birds is observed. Birds of the same kind move together.
(e) Several pairs of objects are collected and the pairs are than paired. Each time there are four objects placed together.

**2.** (a) Several adults are measured. Measuring to the nearest inch they could be under 5′, 5′, 5′1″, 5′2″, . . . , 6′3″, 6′4″, 6′5″, over 6′5″.
(c) A half cup of unpopped corn is popped. {pops, does not pop}.
(e) Several matchbooks are dropped. {front up, front down, on a side}.

**3.** (a) 1. Should be {heart, club, spade, diamond}. 2. Permissible. 3. Permissible. 4. A diamond is also red, and a spade falls into none of the categories. {red, black} would be permissible.
(c) 1. She may be neither, or she may be both. {in a sorority, not in a sorority} is permissible. 2. Permissible, unless you would like to add ambidextrous. 3. Permissible. 4. A girl could be in all three, or she could be in none of these. {an only child, has only sisters, has only brothers, has both brothers and sisters} is permissible.

**4.** (a) They get wet. They may or may not drown.
(c) It stops spinning. It may land on 1,2,3,4,5,6, or between the numbers.

### Section 2

**2.** Let $P(A)$ mean "the probability of outcome $A$." $P$(two heads)$=1/4$; $P$(one head and one tail) $= 1/2$; $P$(two tails) $= 1/4$.

**4.** (a) 1/13. (b) 1/4. (c) 3/13. (d) 4/13.

**6.** Out of every 10,000 people aged 20, approximately 18 will die within one year. No, it only refers to the collection.

### Section 3

**2.** (a) 1/2, 1/2. (b) 1.

**4.** No.

**6.** The seeds will look yellow and smooth. 9 to 3 to 3 to 1.

### Section 4

**1.** (a) 1/2. (b) 1/2. (c) 1/3. (d) 1/3. (e) 1/3. (f) 2/3. (g) 1/6. (h) 1. (i) 1/3. (j) 2/3. (k) 2/3. (l) 1/6.

**3.** (a) 0.20. (b) 0.35. (c) 0.28. (d) 0.59. (e) 0.41. (f) 1.

**5.** The probability is 1/16 for each type. The probability of a seed being yellow and wrinkled in appearance is 3/16.

### Section 5

**1.** (a) $P_0(Y) = 0.55$, $P_0(G) = 0.45$.
(c) $P_0(Y) = 0.50$, $P_0(G) = 0.50$.

**2.** (a) $P_1(YY) = 0.30$, $P_1(GY) = P_1(YG) = 0.25$, $P_1(GG) = 0.20$.
(c) $P_1(YY) = 0.25$, $P_1(GY) = P_1(YG) = 0.25$, $P_1(GG) = 0.25$.

**3.** 22% will be genetically pure blue-eyed, 28% will be genetically pure brown-eyed, and the rest will have brown eyes and carry a recessive gene for blue. The next generation will be the same.

## CHAPTER 9

### Section 1

**1.** (a) pressing the keys; movement inside typewriter; printing the characters.
(b) spoken word received by ear; operations of the brain on the information received; verbal answer given.
(c) lifting the receiver and dialing; telephone connection made; phone rings.
(d) setting the alarm to ring and selecting the time; works move until appropriate time; alarm rings.
(e) coin inserted and drink selector pushed; machine works; drink is dispensed.
(f) dialing the combination; appropriate tumblers move; lock opens.

(g) turning on, selecting channel, and tuning; signal received by set; picture shown on screen.

(h) heat applied to thermometer; mercury expands; length indicates temperature.

3. *Analog:* protractor, sundial, ruler. *Digital:* tallying, cash register, abacus, clock, counter, poker chips, adding machine.

## Section 2

1.

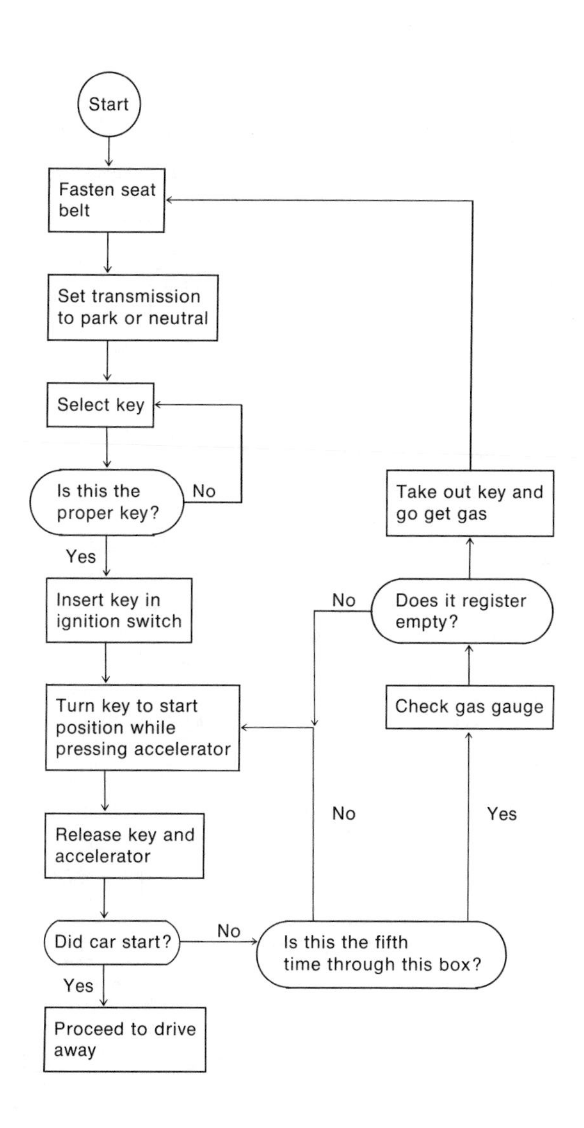

3.

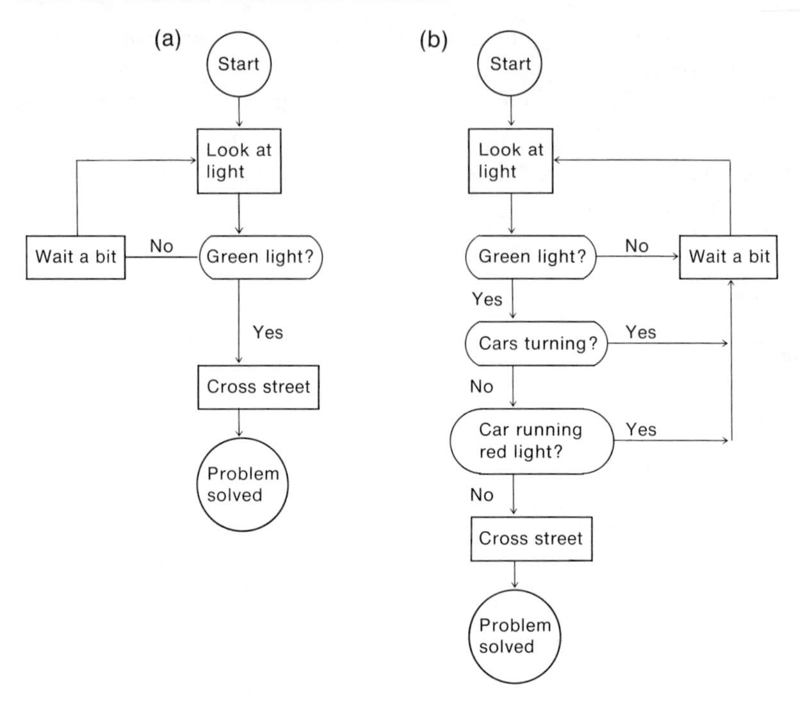
(a)
Start
Look at light
Wait a bit
No
Green light?
Yes
Cross street
Problem solved
(b)
Start
Look at light
Green light?
No
Wait a bit
Yes
Cars turning?
Yes
No
Car running red light?
Yes
No
Cross street
Problem solved

5.

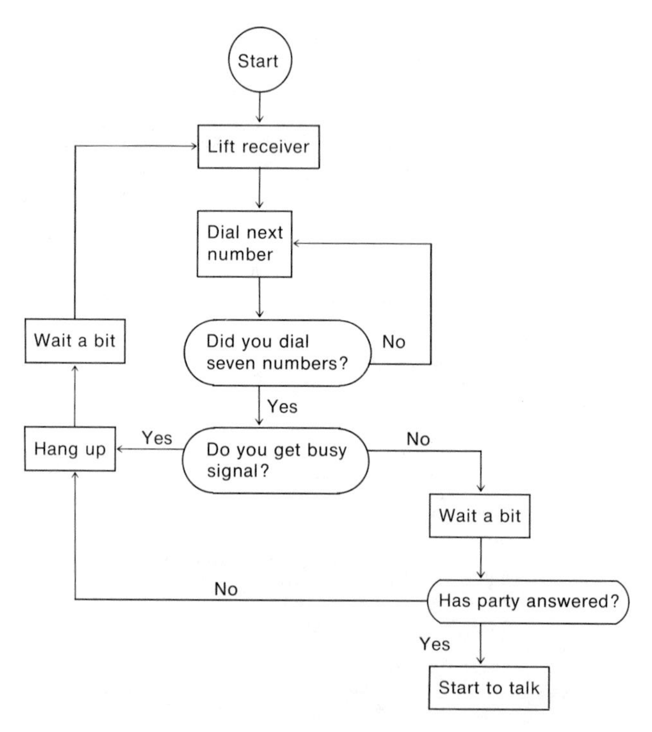
Start
Lift receiver
Dial next number
Wait a bit
Did you dial seven numbers?
No
Yes
Hang up
Yes
Do you get busy signal?
No
Wait a bit
No
Has party answered?
Yes
Start to talk

## Section 3

**1.** 10, −5, 6, 11, 0.

**3.** (a) 11, (b) 36, (c) 43.

**9.** Some provision must be made for the sign of the quotient.

## Section 4

**1.** With rounding to two significant digits, repeated addition gives 900; repeated addition with shift, 870; actual value, 871.

**3.** Evaluating as written with rounding: 34.80. Evaluating $x[x(6x - 2) - 4] + 1$ with rounding: 34.84. Actual value: 34.830762.

# Index